T0136798

Physics Demonstrations

A Sourcebook for Teachers of Physics

Julien Clinton Sprott

The University of Wisconsin Press

The University of Wisconsin Press
1930 Monroe Street, 3rd Floor
Madison, Wisconsin 53711-2059
uwpress.wisc.edu

3 Henrietta Street, Covent Garden
London WC2E 8LU, United Kingdom
eurospanbookstore.com

Printed in the United States of America

Library of Congress Cataloging-in-Publication Data

Sprott, Julien C., author.
Physics demonstrations : a sourcebook for teachers of physics / Julien Clinton Sprott.
pages cm
Includes bibliographical references and index.
ISBN 978-0-299-30470-6 (cloth : alk. paper) — ISBN 978-0-299-30473-7 (e-book)
1. Physics—Experiments. 2. Physics—Study and teaching. I. Title.
QC33.S675 2015
530.078--dc23

2014035272

Disclaimer

The demonstrations and other descriptions of procedures and use of equipment in this book have been compiled from sources believed to be reliable and to represent the best opinion on the subject as of 2005. However, no warranty, guarantee, or representation is made by the author as to the correctness or sufficiency of any information herein. The author does not assume any responsibility or liability for the use of the information herein, nor can it be assumed that all necessary warnings and precautionary measures are contained in this publication. Other or additional information or measures may be required or desirable because of particular or exceptional conditions or circumstances, or because of new or changed legislation. Teachers and demonstrators must develop and follow procedures for the safe performance of the demonstrations in accordance with local regulations and requirements.[1]

[1] See the essay "Injuries in School/College Laboratories in the USA" by Dr. Ronald B. Standler at http://www.rbs2.com/labinj2.pdf, which discusses the law in the United States about injuries to pupils or students in school or college science laboratories.

Contents

Preface

In the matter of physics, the first lessons should contain nothing but what is experimental and interesting to see. A pretty experiment is in itself often more valuable than twenty formulae extracted from our minds.[1]

— *Albert Einstein*

Demonstrations greatly enhance the teaching of physics. Visual examples of abstract concepts contribute immeasurably to their mastery. They also provide an opportunity to illustrate the scientific method and to teach the student to relate experimental observation to scientific theory. Experiments represent the means by which scientific knowledge has advanced so rapidly in modern times. Finally, not to be underestimated, the use of demonstrations provides a powerful motivation and makes the learning of physics much more enjoyable!

This book is a compilation of many of the demonstrations that have been used at the University of Wisconsin – Madison in the teaching of elementary physics over the years as well as a number of demonstrations and embellishments I have developed for use in a series of popular lectures, "The Wonders of Physics," aimed at a public audience and children in particular. Although I can claim very little credit for developing most of the demonstrations in this book, I have done them all personally and have tested them in a classroom setting or before an audience of the general public.

To write a book that purports to describe all possible physics demonstrations would be a formidable task since the discipline of physics is so all encompassing. Instead, I have selected demonstrations that are especially dramatic or provocative, as well as some of the standard demonstrations that can be presented in unusual ways. In part, this reflects my feeling that it is necessary to get the attention of the students or the more general audience and convince them that physics is interesting before any real learning can occur, but also because there seems to be little help available for the instructor who wants to improve the quality of his or her presentation or adapt familiar demonstrations for the public. I do not intend for this book to contain a complete set of demonstrations for the teaching of elementary physics, but rather to provide the instructor with ideas for new demonstrations and more engaging ways to present old ones.

---

[1] A. Moszkowski, *Conversations with Einstein*, Horizon Press: New York (1970), page 67.

There are many other books on physics demonstrations listed in the bibliography. A number of these are far more complete than the book I have written. Some are quite old and do not make use of the advances in instrumentation of the past few decades. I have attempted to complement these works by concentrating less on the description of the apparatus, much of which is now well known and commercially available, and more on techniques for bringing them to life and generating interest in them. For example, I have included historical anecdotes and other commentary to add to the interest and have suggested alternate and unusual uses for the apparatus. I have provided brief explanations of those facets of the demonstrations that may not be obvious to a trained physicist or physics teacher. I have also emphasized safety more than is usual in such books.

Most books on physics demonstrations describe the important concepts and then give a number of demonstrations to illustrate each one. The majority of these demonstrations, though pedagogically sound, are not very suitable because they require close, careful measurement, or they are not easily visible to a large audience, or they are simply not very exciting. They belong in a laboratory, not as part of a lecture. I have taken a different approach, in that I describe the demonstrations first and then mention the many physics concepts that are illustrated. I have found this departure from the usual teaching mode in which one introduces the concepts sequentially to be especially effective for informal presentations to the public and to children who have not studied physics, but I also use it to motivate students in conventional physics courses.

When I cause a soap bubble filled with methane gas to explode by touching a lighted candle to it (section 2.21), no one in the audience is asleep, and I can proceed to discuss thin films, surface tension, Archimedes' principle, or any number of other related topics with their full attention. There are certainly more "scientific" ways to demonstrate the variation of the speed of sound with the density of a gas than by breathing helium and sulfur hexafluoride (section 3.3), but few students will forget the result, and it is hard to resist wanting to know the explanation. The talking head (section 6.9) and Pepper's ghost (section 6.10) are extraordinarily elaborate demonstrations of the rather simple ideas of the total and partial reflection of light off the interface between two media, but when you contrast this method with the way such concepts are usually taught, it is no wonder why physics has the reputation of being dry and uninteresting.

It is my hope that this book and the accompanying video clips available at http://physicsdemonstrationsvideos.com/ will also inspire teachers of physics to take the demonstrations beyond the classroom to the general public. All that is required is a notice to the media and a carefully planned, fast-paced presentation of twenty to thirty demonstrations. Such a program never fails to attract an enthusiastic audience. The public is hungry for entertainment with educational value, and demonstrations such as described in this book will appeal to adults and children alike.

My own program is a yearly event with the best of the old demonstrations and a few new ones each year. I repeat the program ten times during two successive weekends in February to accommodate the demand. This format is preferable to using a larger hall in which it might be difficult to see many of the demonstrations. The schedule avoids saturating the interest and allows adequate time to develop dramatic new demonstrations. The demonstrations for the general public generate a continual demand for special presentations to schools and other groups, which we accommodate through a traveling version of the presentation that uses smaller, more portable variants of many of the demonstrations described here. "The Wonders of Physics" program is described more fully at http://wonders.physics.wisc.edu/. You will also find there links to several dozen hour-long free videos of the yearly presentations dating back to 1986.

This book assumes you have some knowledge of physics, and thus I have not explained every scientific principle in complete detail. Many elementary physics texts are available to fill that need. Instead, I have emphasized those aspects of the physics that might be unfamiliar or overlooked as well as ways in which you could introduce and explain the concepts effectively to an audience of the general public.

The apparatus required for the demonstrations varies considerably in complexity. You can do some of the demonstrations with equipment available at home or from a local hardware store. Others are best constructed from components found in a well-equipped physics laboratory or available from specialized mail-order vendors. In such cases, I have tried to provide references to articles that describe the construction or have outlined the important considerations that must go into the design, but I have largely avoided a cookbook approach to the assembly of the required apparatus since there is usually great leeway in the design depending on the resources available. Some of the more complicated demonstrations are available ready-made from vendors of scientific apparatus, and in such cases I have included possible sources in a footnote.

Although most of the demonstrations are completely safe, a few involve potential dangers and require a good knowledge of physics as well as experimental technique. The dangers include burns from heat, caustic chemicals, and high-frequency electricity; frostbite from handling cryogenic materials; cuts, especially from broken glassware; explosions from high-pressure gases and volatile substances; implosions from evacuated glassware; asphyxiation; electrocution; eye damage from lasers; and exposure to radioactivity. I have mentioned the major hazards, but a large dose of caution and a respect for the forces of nature are essential.

I am indebted to many people in the preparation of this work. I have patterned the book after *Chemical Demonstrations* by Professor Bassam Shakhashiri, whose example and encouragement also inspired "The Wonders of Physics" lectures. Many of the demonstrations have been handed down within the University of Wisconsin – Madison Physics Department for so many years that their origin is unknown. Others have come from the suggestions of friends and colleagues too numerous to mention. To all who have contributed, I offer my sincere thanks.

Julien Clinton Sprott
Emeritus Professor of Physics
University of Wisconsin – Madison
Madison, Wisconsin
September 2014

Acknowledgments

Durlin C. Cox, Resonance Research, Inc., Baraboo, WI

Roger E. Feeley, University of Maine, Orono, ME

Robert G. Greenler, University of Wisconsin, Milwaukee, WI

Donald W. Kerst, University of Wisconsin, Madison, WI

Cassandra E. Kight, University of Wisconsin, Madison, WI

Lawrence S. Lerner, California State University, Long Beach, CA

Thomas W. Lovell, University of Wisconsin, Madison, WI

Kenneth C. Maas, University of Wisconsin, Madison, WI

Edward E. Miller, University of Wisconsin, Madison, WI

Mary C. Moebius, Verona Middle School, Verona, WI

Steve R. Narf, University of Wisconsin, Madison, WI

David E. Newman, University of Alaska, Fairbanks, AK

Paul D. Nonn, University of Wisconsin, Madison, WI

Steve P. Oliva, University of Wisconsin, Madison, WI

James C. Reardon, University of Wisconsin, Madison, WI

Stephen M. Salemson, University of Wisconsin, Madison, WI

Bassam Z. Shakhashiri, University of Wisconsin, Madison, WI

Elizabeth A. Steinberg, University of Wisconsin, Madison, WI

Mark A. Thomas, University of Wisconsin, Madison, WI

Teri L. Vierima, Resource Strategies, Inc., Madison, WI

Christopher Watts, New Mexico Tech, Socorro, NM

William P. Zimmerman, University of Wisconsin, Madison, WI

1
Motion

Motion, more properly called "mechanics," is the oldest branch of physics, having been put on a firm quantitative basis by Isaac Newton[1] (1642 – 1727), who by the age of 24 had also developed calculus (or "fluxions," as he called it), which subsequently became an indispensable tool of science. Many people consider the study of motion as relatively mundane, but Albert Einstein[2] (1879 – 1955) at the age of 26, while employed at the Swiss Patent Office, thought deeply about the motion of light and revolutionized our understanding of the relation of space and time. More recently, the discovery and understanding of deterministic chaos in simple nonlinear dynamical systems has rejuvenated the interest in classical mechanics. The study of motion offers the opportunity to develop concepts such as the conservation of energy that are relevant to all branches of physics. Motion demonstrations illustrate these concepts in their least abstract and most easily visualized form.

---

[1] Newton was an English physicist and mathematician, who, in addition to developing classical mechanics and gravitation, invented calculus (along with Gottfried Wilhelm Leibnitz, 1646 – 1716) and showed that white light contains the colors of the rainbow, but he was entirely without humor and retired from research, spending the last third of this life as a government official following a nervous breakdown in 1693.

[2] Einstein was rather slow as a student, and his teachers thought he would never amount to much. In later life, he was known for not wearing socks or a tie and for rarely getting a haircut.

1.1
Guinea and Feather Tube

In an evacuated glass tube, objects fall at the same rate independent of their size, shape, and mass.

MATERIALS

- cylindrical glass tube with removable end and pump-out nozzle[1]
- penny or other small dense object
- feather or ball of cotton
- rubber ball
- vacuum pump
- fluorescent paint and ultraviolet lamp (optional)
- pressure gauge (optional)
- heavy book and sheet of paper (optional)

PROCEDURE

Equip a glass tube at least a meter in length with a valve and nozzle through which you can evacuate the tube with a mechanical vacuum pump. Such a tube is called a "guinea and feather tube" because those were the objects traditionally used when performing the demonstration in England centuries ago. The guinea was an English gold coin issued from 1663 to 1813 and valued at 21 shillings. These days, a modern coin and a ball of cotton are normally used.

Introduce the demonstration by asking the audience which falls faster, a feather or a coin. (The question is reminiscent of the old joke about which weighs more, a pound of

[1] Available from Carolina Biological Supply Company, Frey Scientific, PASCO Scientific, Sargent-Welch, and Ward's Science.

feathers or a pound of lead.) Whichever answer is given is either right or wrong depending on whether the experiment is performed in the air or in a vacuum. Point out that science often suggests opposing theories, and in such cases, experiments are required to determine which theories are correct. An experiment cannot prove that a theory is correct, only that it is incorrect. Point out that truth in science is not determined by voting or by the strength of authority, but rather by experiment.

Begin by standing on top of the lecture bench with two objects, one significantly heavier than the other, such as a rubber ball and a similar sized ball of cotton. Ask the audience which will hit the floor first if you drop them simultaneously. Many people will say they will hit at the same time, having heard that before or having some knowledge of physics. Drop them to show that the heavier one does hit first. Then explain that the reason is that air resistance slows the lighter object more than the heavier object.

Then offer to repeat the demonstration in an evacuated tube. Place the two objects in the tube at atmospheric pressure and rapidly invert the tube, yielding the expected result. Then evacuate the tube and ask the audience what is in the tube. Correct answers are "a penny," "a ball of cotton," "a partial vacuum" or "a little bit of air." Repeat the demonstration to show that the two will fall at the same rate when you evacuate the tube. Finally, let air back into the tube, and show that the cotton again falls slowly. If it is difficult to see the penny, ask the audience to listen for it to hit the bottom of the tube. You can put fluorescent paint on the ball of cotton and illuminate the tube with an ultraviolet lamp for better visibility. To save time, you could evacuate the tube before doing the demonstration.

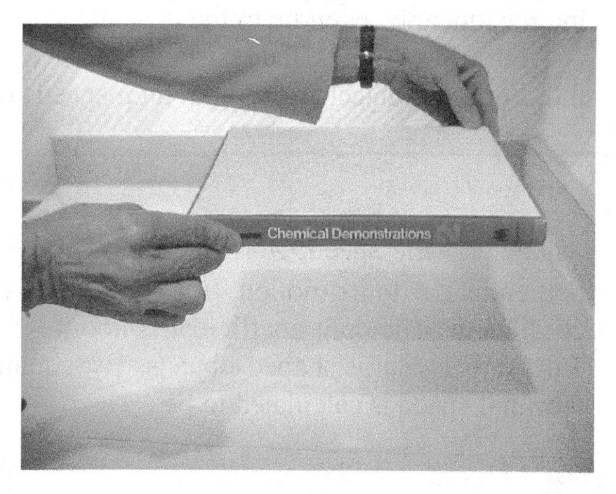

If a vacuum pump is not available, you can do an alternate form of the demonstration using a heavy book and a sheet of paper (smaller than the book) [1]. First drop the paper and book side-by-side. The book will fall much faster. Then place the paper flat beneath the book and release the two together. They will fall at the same rate, but many people will assume the reason is that the book pushes down on the paper. Then place the paper flat against the top of the book, being sure it does not extend over the edges of the book. They will again fall at the same rate because the book eliminates the air resistance that the paper would otherwise experience. Finally, crumple the paper into a tight wad and drop it again.

DISCUSSION

Legend has it that Galileo Galilei (1564 – 1642) performed this experiment by dropping two cannonballs, one ten times heavier than the other, from the Leaning Tower of Pisa [2]. The legend is almost certainly false, although Galileo did perform similar experiments in his youth, such as rolling balls of different masses down an inclined plane [3, 4], primarily because he lacked timing devices of sufficient accuracy to study freely falling bodies. He realized that when dropped from a sufficient height, the heavier mass would indeed fall faster [5]. However, his arguments relied more on thought experiments, such as whether two identical objects connected together would fall at the same rate as they would separately, rather than on actual experiments. On August 2, 1971, astronaut David R. Scott repeated the experiment with a geologist's hammer and a falcon's feather while standing on the airless surface of the moon while the world watched on television.

The fact that a light object falls as fast as a heavy object in a vacuum puzzles many people who correctly reason that gravity should pull harder on the heavy object. However, from Newton's second law of motion ($F = ma$), a harder pull is required to accelerate a heavier object, and the effects just cancel. This demonstration illustrates the equivalence of gravitational and inertial mass and is the basis of the equivalence principle upon which the general theory of relativity is based [6]. In 1890, the Hungarian physicist Baron Lorand von Eötvös (1848 – 1919) refined Galileo's experiment and showed that inertial and gravitational mass are the same to one part in 10^8, and more recent experiments [7], motivated by the search for a new form of force [8], have reached five parts in 10^{10}.

The concept of mass was introduced by Aegidius Romanus (ca. 1243 – 1316), and the concept of inertial mass was introduced by Johannes Kepler (1571 – 1630), but the distinction between mass and weight remained in a confused state until Jean Richer (1630 – 1696) inadvertently discovered (in 1671) that weight varied with location on the Earth. Isaac Newton (1642 – 1727) explained this effect through his universal law of gravitation, although people continued to use the terms "weight" and "mass" somewhat interchangeably, even up to the present.

For most objects, the resistance of the air noticeably alters the speed with which they fall. For large objects moving at high speeds through air, the drag force is

$$F_d = C\rho A v^2/2$$

where v is the speed, A is the cross-sectional area of the object measured in a plane perpendicular to its motion, ρ is the density of the air, and C is the (dimensionless) drag coefficient. The drag coefficient is about 0.5 for spherical objects but can be as high as 1.0 for irregularly shaped objects. By equating the drag to the weight ($F_d = mg$), the terminal speed is calculated to be

$$v_t = \sqrt{\frac{2mg}{C\rho A}}$$

For two equally shaped objects, the one falling faster will have the greater drag force, although the force is never sufficient to overcome its greater weight and make the less massive object fall faster [9]. The buoyant force of the air contributes a small, additional, upward force on the object, especially if the object has a small mass density.

HAZARDS

A significant hazard of this demonstration is breakage of the glass tube, especially if it strikes a hard object when you rapidly invert it while evacuated. You should use strong glass and take great care during handling. You should provide some sort of cradle to prevent the tube from rolling off the table and falling onto the floor when not in use. An overly heavy object inside the tube could also break the tube when it falls. Make sure the belt of the vacuum pump has a guard, or place the pump out of reach.

REFERENCES

1. T. L. Liem, *Invitations to Science Inquiry*, Ginn Press: Lexington, MA (1981).
2. L. Cooper, *Aristotle, Galileo and the Tower of Pisa*, Cornell University Press: Ithaca, NY (1935).
3. A. R. Hall, *From Galileo to Newton, 1630 – 1720*, Harper and Row: New York (1963).
4. S. Drake, *Scientific American* **238**, 84 (May 1973).
5. H. Butterfield, *The Origins of Modern Science*, Macmillan: New York (1960).
6. D. Sciame, *Scientific American* **196**, 99 (Feb 1957).
7. T. M. Niebauer, M. P. McHugh, and J. E. Faller, *Phys. Rev. Lett.* **59**, 609 (1987).
8. E. Fischbach, D. Sudarsky, A. Szafer, C. Talmadge, and S. H. Aronson, *Phys. Rev. Lett.* **56**, 3 (1986).
9. R. Weinstock, *Phys. Teach.* **31**, 56 (1993).

1.2 Reaction Time

The distance an object falls in the gravitational field of the Earth provides a sensitive measure of short time intervals.

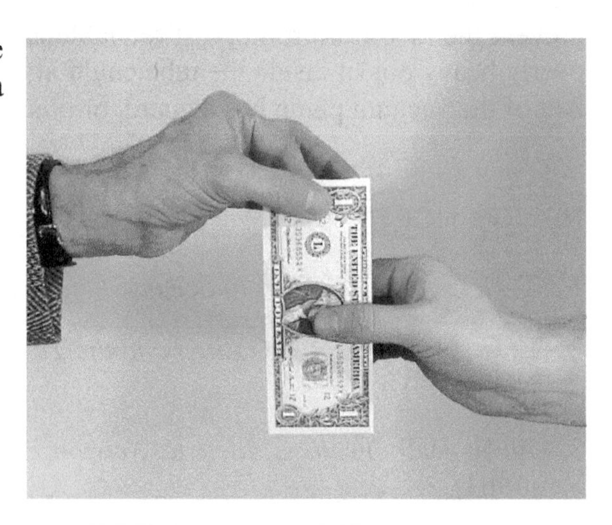

MATERIALS

- meter stick
- crisp dollar bill

PROCEDURE

The usual demonstration consists of holding a meter stick vertically from the top while a volunteer stands poised ready to catch it between the thumb and forefinger [1]. If the fingers are opposite the 50-cm mark, for example, when you drop the meter stick, the position of the fingers when the meter stick is caught gives a measure of the distance the meter stick fell before the volunteer could react. The time is then

$$t = \sqrt{\frac{2d}{g}}$$

where $g = 9.8$ m/s^2 and d is the distance dropped (in meters). For extra interest, do the demonstration using a crisp dollar bill with the volunteer's thumb and forefinger opposite George Washington's portrait. You can fold the bill lengthwise to ensure that it drops straight down not deflected by the air. Only a small fraction of the volunteers are able to catch the bill. Without much risk, you can offer the bill to the volunteer if he or she catches the bill. Sometimes the volunteer will anticipate when you are about to release the bill. You can flinch first to see if they are anticipating. If they grab the bill while it is still in your hand, you can reprimand the volunteer and then drop it while still talking.

Encourage the audience to try this experiment at home, perhaps using a yardstick, and to compare their reaction time with that of their friends and relatives. Ask them to compare the reaction times of people of different ages and to report back with the results.

In addition to illustrating the motion of an object under constant acceleration, you can use such a demonstration to discuss the collection and analysis of scientific data. By performing the experiment with a number of volunteers and tabulating the results, you can explain the concepts of mean, average, and standard deviation. You can plot the results as a histogram. You can tell the fastest volunteers, in the tail of the distribution, to become racecar drivers, although reaction time is only one quality needed for such a profession.

Whenever you use a volunteer from the audience, take a few moments to become acquainted with the person. Ask the person's name, whether they like science, what they study, what they want to be when they get older, and something relevant to the demonstration in which they are about to participate. In a case like this, ask if they like having money and if they would like to earn a dollar or whether they think they have quick reactions. Ask whose portrait is on the dollar bill (George Washington), who he was, and whether he is still president.

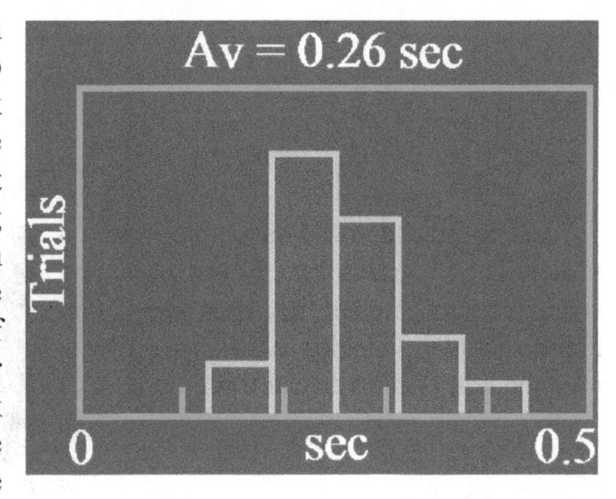

DISCUSSION

The distance from the center of a dollar bill to the end is about seven centimeters. The time it takes something to fall seven centimeters is about a tenth of a second. Most people's reaction time is a few tenths of a second. Therefore, unless they anticipate that you are about to drop the bill, they will not react quickly enough.

The time required for an object to fall provides a practical way to measure, say, the depth of a well. Drop a rock into the well and measure the time required for it to reach the bottom. The depth in feet is $16t^2$ where t is the time in seconds. You can determine the height of a cliff in a similar manner by throwing the rock horizontally off the edge of the cliff. However, if the height is too great, the calculated height will be too large because air resistance limits the speed of the falling object to its terminal speed.

The frictional force for an object moving through the air is given approximately by $f = CAv^2$, where v is its velocity, A is its cross-sectional area, and C is the coefficient of air resistance, which has a typical value on the order of 1 kg/m^3. Equating this force to the weight of the object gives the terminal speed, which is about 140 miles/hr (or 60 m/s) for a skydiver, depending on the orientation of the body.

HAZARDS

There are no hazards with this demonstration other than to the ego of the slowest volunteers. Do not drop or throw rocks if there is any danger to people or property below.

REFERENCE

1. H. F. Meiners, *Physics Demonstration Experiments*, Vol I, The Ronald Press Company: New York (1970).

1.3 Ballistics Car

A car rolling across the table fires a projectile straight upward and subsequently catches it, illustrating that the horizontal velocity of a projectile is independent of the vertical force on it.

MATERIALS

- car with spring-loaded mechanism[1]
- steel ball
- cardboard or other tunnel (optional)
- inclined plane (optional)
- piece of chalk (optional)

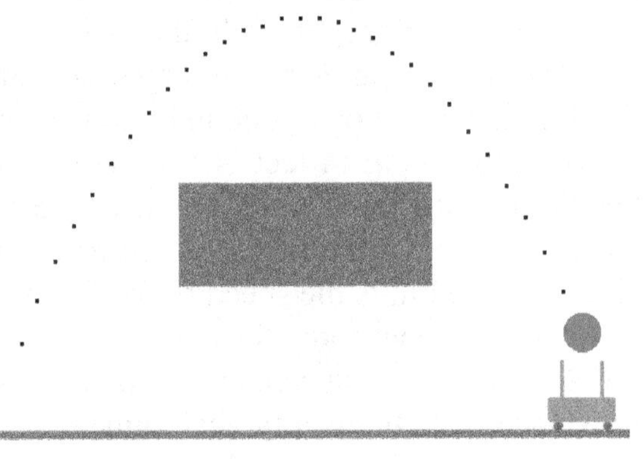

PROCEDURE

A small car that rolls across a table is equipped with a spring-loaded mechanism that propels a steel ball vertically upward (also called a "Howitzer"). As you roll the car across the table, the string, which is attached securely to something on the table, becomes taut and pulls out a pin, releasing the spring without significantly altering the velocity of the car. The car catches the ball as it falls. Use an air hose or a board and eraser as a bumper to prevent the cart from rolling off the bench.

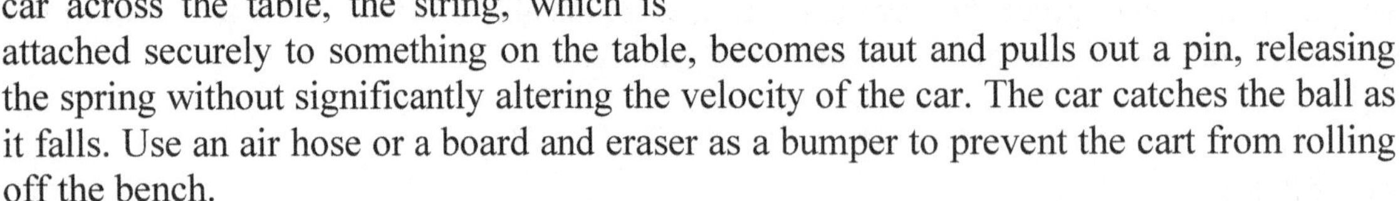

First explain the operation of the device to the audience, and pull the pin out while the car is at rest. Then roll the car across the table to complete the demonstration. As an embellishment, make the car go through a short tunnel while the ball goes over the top of the tunnel. In a larger scale demonstration, use a car large enough to accommodate a rider [1].

In a variation of the demonstration, attach the car to a horizontal string that passes over a pulley and supports a weight that accelerates the car. Ask the audience to vote on where the ball will fall relative to the car before performing the demonstration. Most people correctly surmise that it lands behind the car. Another variation is to let the car

[1] Available from Carolina Biological Supply Company, Frey Scientific, Sargent-Welch, Science First, and Science Kit & Boreal Laboratories.

roll down an inclined plane [2]. Even physics teachers often will not correctly answer that the ball remains above and lands in the car provided the moment of inertia of the wheels is negligible [3].

If you do not have the apparatus needed to do this demonstration, you can illustrate the same principle by walking quickly but steadily across the room while throwing a piece of chalk vertically upward, catching it in your hand [4].

DISCUSSION

This demonstration illustrates the independence of the vertical and horizontal motion of projectiles. Since the ball launches from a car moving with a constant horizontal velocity, the ball has the same horizontal velocity as the car and thus remains directly above it (ignoring friction of the wheels and the possibly different air resistance of the ball and the car). Alternately, you can use this demonstration to illustrate the transformation of coordinate systems. In the inertial system moving with the car, the ball moves straight up and down. In the system fixed in the room, the ball follows a parabolic trajectory.

HAZARDS

You should take care that the car does not roll off the edge of the table, not so much because it might injure someone, but because the car could be damaged if it fell on the floor. Also keep the area above the car clear to prevent the ball from hitting anything. Be sure to retrieve the ball if it ends up on the floor so that someone does not step on it and slip.

REFERENCES

1. H. F. Meiners, *Physics Demonstration Experiments*, Vol I, The Ronald Press Company: New York (1970).

2. R. Prigo and A. Rosales, *Am. J. Phys.* **44**, 783 (1976).

3. R. A. Serway, J. Lehman, and R. Hall, *Phys. Teach.* **33**, 578 (1995).

4. E. van den Berg and R. van den Berg, *Phys. Teach.* **36**, 356 (1998).

1.4
The Monkey and the Coconut

A projectile aimed at a stuffed monkey hits the monkey despite the fact that the monkey begins to fall at the instant the projectile is fired.

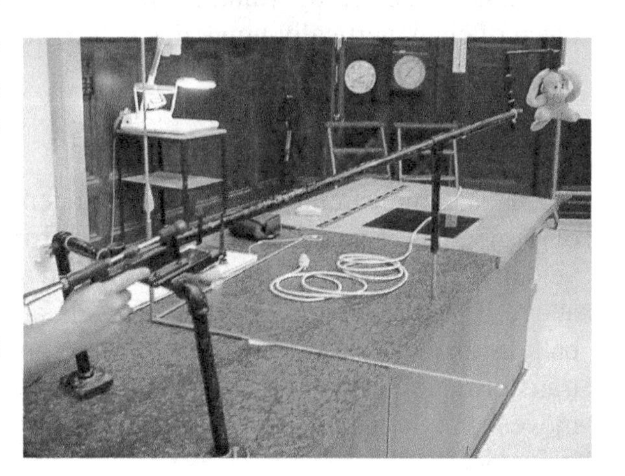

MATERIALS

- spring-loaded or pneumatic gun with projectile[1]
- stuffed monkey
- electromagnet
- micro switch or optical sensor

PROCEDURE

This popular demonstration usually goes by the name "The Monkey and the Hunter," but is here modified to appeal to the sensitivity of those ethically opposed to killing monkeys. You can explain that this is a circus monkey, and give him a name such as "Darwin." Ask the audience to consider tossing a coconut to a monkey who drops from a tree at the instant the coconut is thrown and who tries to catch the coconut. If the coconut is thrown directly at the monkey in the tree, you might expect it to go above the monkey since the monkey is falling. On the other hand, the coconut does not follow a straight line.

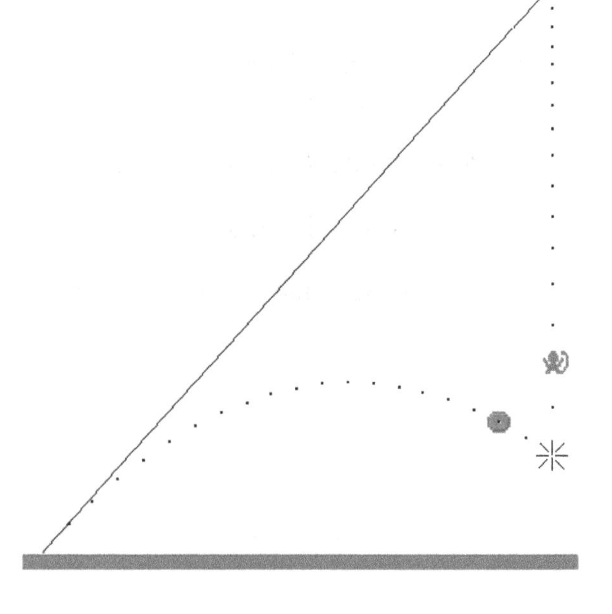

Ask the audience to vote on whether a coconut aimed directly at the monkey will go above or below the monkey as he falls. Those who abstain from voting are of course correct because the distance fallen by the monkey and the coconut are the same (neglecting air resistance) during the time the coconut is in flight. You can best

[1] Available from Fisher Science Education, PASCO Scientific, and Sargent-Welch.

illustrate the effect by drawing a straight line to the position of the monkey in the tree. The coconut falls below this line by the same distance the monkey drops in a given time. The problem is slightly whimsical because the monkey is not likely to react instantly to the release of the coconut, and air resistance acts differently on the monkey and coconut.

The apparatus for this demonstration usually consists of some form of spring-loaded or pneumatic gun that electrically deactivates an electromagnet that supports a stuffed monkey [1–4]. Alternately, you can fire the gun mechanically and make it de-energize the magnet by a micro switch or optical sensor. A useful embellishment is the ability to change the vertical angle between the gun and the monkey and the ability to change the speed of the projectile. Neither should have an effect on the outcome of the demonstration. The effect is most dramatic if there is a considerable separation between the gun and the monkey, but the precision with which the apparatus is constructed determines the allowable separation. It is best to hit the monkey every time even if you have to shorten the distance. You can also use a dart for the projectile, which will impale the monkey, but this variant is more dangerous. For extra drama, have someone barge into the room in a gorilla suit or monkey mask to rescue the monkey off the floor and carry it out of the room, perhaps to be returned later in some fashion.

DISCUSSION

This demonstration illustrates the fact that the vertical and horizontal components are independent in projectile motion. The vertical acceleration of an object is independent of both its horizontal and its vertical velocity. Another common example of this idea is the problem of where to release an object from a moving airplane in order to hit a target on the ground and then to find where the airplane is in relation to the target when it is hit.

Note that a collision will occur only if $v_o \sin \theta_o > \sqrt{gd/2}$, where v_o is the initial speed of the coconut, θ_o is its initial angle with respect to the horizontal, d is the initial elevation of the monkey above the ground, and g is the acceleration due to gravity (9.8 m/s^2). If $v_o \sin \theta_o$ is less than this value, the monkey cannot catch the coconut because the coconut will strike the ground before reaching him. Also note that the horizontal component of momentum is conserved when the coconut strikes the monkey, causing the monkey and coconut to continue moving in the direction of the coconut but with a smaller velocity.

Those who have experience with guns will know that guns are "sighted in" at a given distance (100 meters, for example) to account for the downward displacement of the bullet during its travel. Thus when a fixed target is closer than the distance for which the gun was sighted, it is necessary to aim below the target in order to hit it [5]. In such a case, with a falling target, you would always have to aim below it.

HAZARDS

The danger in this demonstration is that the projectile will hit someone in the audience either directly or after deflecting off something along the path of its trajectory. Take care in aiming the gun, and move any objects that could deflect the projectile. Use a soft monkey and projectile.

1.5
Inertia Ball

The property of inertial mass is illustrated by pulling on a string attached to the bottom of a heavy ball that is suspended by an identical string until one of the strings breaks.

MATERIALS

- heavy iron ball with hooks on opposite sides (second ball optional)[1]
- several lengths of string
- rigid support stand
- rubber pad
- cart on inclined plane with two identical rubber bands (optional)
- heavy concrete block and sledge hammer (optional)

PROCEDURE

Suspend two identical heavy steel balls with hooks on opposite sides from a support using strings [1–8]. It is important for the support to be quite rigid. A string dangles from the bottom of each ball. Tell the audience that all the strings are identical and ask them to vote on which string will break first if you pull down on the lower string on one of the balls. Many people will assume the upper string will break first. Then give the lower string a jerk to make it break below the ball.

While the spectators are puzzling over why their intuition was wrong, offer to repeat the experiment with the second ball while commenting on the importance of repeating scientific experiments, but this time pull the string gently to make the string break above the ball. Ask the audience to explain why the result was different. If the vote were the opposite, reverse the order of the demonstrations. Discuss the importance of controlling

[1] Available from Carolina Biological Supply Company, Fisher Science Education, Sargent-Welch, Science First, and Ward's Science.

all the relevant variables in an experiment. You can use a single ball if you take the time to retie the string between each demonstration or if the ball has two strings attached to its bottom and you break them in the right order. Another variant of the demonstration is to lift the ball slowly with the string and then try to lift it with a quick jerk, breaking the string. In this variant, you can use a paint can filled with sand in place of the ball.

Many people misunderstand this demonstration, thinking that with the quick jerk, the force does not have time to propagate to the upper string, but that would violate Newton's third law of motion. A more instructive demonstration involves a heavy cart on an inclined plane, held in place with a rubber band, with a second identical rubber band connected to the downhill side of the cart. If you pull the second rubber band slowly, the upper rubber band will break, but if you pull it quickly, the lower one will break. This demonstration illustrates that it is the amount of stretch that determines which band breaks, which is less obvious with the conventional demonstration using an inertia ball.

A more dramatic popular demonstration of the same law of inertia uses a heavy concrete block placed on the chest of someone lying down, perhaps even on a bed of nails, while the block is broken with a sledgehammer without injuring the person [9–11]. This demonstration is not recommended, because of the risk to the person if the hammer misses the block or if fragments from the broken block hit someone. Another popular inertia demonstration involves breaking a broomstick supported on either end with delicate wine glasses by striking the stick in its middle with another stout stick [12].

DISCUSSION

The explanation involves the inertia of the ball. With a quick jerk, the ball has to accelerate, and a considerable force is required to do this if the mass of the ball is large ($F = ma$). On the other hand, with a slow pull, the acceleration is negligible, and the upper string is supporting the weight of the ball plus the tension in the lower string, causing the upper string to break. From Newton's second law of motion,

$$T_u - T_l = m(g - a)$$

where T_u is the tension in the upper string, T_l is the tension in the lower string, m is the mass of the ball, g is the acceleration due to gravity (9.8 m/s^2), and a is the downward acceleration of the ball. Thus the upper string breaks when the downward acceleration of the ball is less than 9.8 m/s^2; otherwise the lower string breaks. Although the acceleration of the ball may be large, it is sufficiently brief that there is no noticeable displacement of the ball.

HAZARDS

The danger is that, when the upper string breaks, the ball could fall on your hand or foot. Use a rubber pad underneath to catch the ball, and remove your hand quickly when the string breaks.

REFERENCES

1. R. M. Sutton, *Demonstration Experiments in Physics*, McGraw-Hill: New York (1938).
2. P. LeCorbeiller, *Am. J. Phys*. **45**, 156 (1945).
3. P. LeCorbeiller, *Am. J. Phys*. **14**, 64 (1946).
4. J. S. Miller, *Physics Fun and Demonstrations*, Central Scientific Company: Chicago (1974).
5. F. G. Karioris, *Am. J. Phys*. **46**, 710 (1978).
6. G. D. Freier and F. J. Anderson, *A Demonstration Handbook for Physics* (2nd ed.), American Association of Physics Teachers: College Park, MD (1981).
7. M. A. Heald and G. M. Caplan, *Phys. Teach*. **34**, 504 (1996).
8. G. M. Caplan and M. A. Heald, *Am. J. Phys*. **72**, 860 (2004).
9. G. L. Hodgson, *Phys. Teach*. **13**, 52 (1975).
10. M. Bucher, *Am. J. Phys*. **56**, 806 (1988).
11. D. P. Taylor, *Phys. Teach*. **34**, 227 (1996).
12. K. C. Mamola and J. T. Pollock, *Phys. Teach*. **31**, 230 (1993).

1.6
Beaker and Tablecloth

A glass beaker, partially filled with colored water, rests near the edge of a table on a cloth that you rapidly pull out from underneath the beaker without spilling the water or breaking the beaker, illustrating Newton's first law of motion.

MATERIALS

- large glass beaker with smooth bottom or a wine glass
- smooth cloth without seam
- water with food coloring
- basketball or other spherical or cylindrical objects (optional)

PROCEDURE

Half fill the beaker with water and place it on the cloth some distance from the edge of the table. You can add a little food coloring to the water to make it more visible. Make sure the cloth, table, and beaker are clean and completely dry. The table and the bottom of the beaker should be smooth, and the cloth should not have a seam. A soft, flexible material such as silk is desirable. Ask the audience whether it is possible to pull the cloth out from under the beaker without spilling the water or breaking the beaker. Many people will say yes, since they have seen this done before. Then slowly pull the cloth until the beaker is a few cm from the edge of the table, and ask the question again. Ask them what you are doing wrong. Many people will say to do it quickly. After looking skeptical, jerk the cloth out from under the beaker, being careful not to pull upward on the cloth. Start with the cloth limp so that your hand can gain some speed before the cloth becomes taut. The beaker should remain on the table, and no water should spill. As you gain confidence, you can do the demonstration with other objects such as dinner plates and drinking glasses, but it is best if the objects have smooth bottom surfaces. You can use a paper towel or sheet of paper instead of the cloth and coins or books instead of the beaker. For added drama, do the demonstration with an entire place setting of dinnerware, including a delicate wine glass filled with wine-colored water [1]. Caution children not to try this at home with their mothers' dishes.

You can also replace the beaker with a basketball or other spherical or cylindrical object [2]. Ask the audience which way the basketball will roll when you pull on the cloth. It will roll forward, eventually coming to rest through friction when it is free of the cloth. Try comparing a hoop with a disk, in which case the hoop will move faster and farther because of its larger moment of inertia relative to its mass. This demonstration is similar to the one in which various objects roll down an inclined plane (see section 1.9).

DISCUSSION

According to Newton's first law of motion (the law of inertia), an object at rest tends to remain at rest until acted upon by an external force. In this case, the external force is the friction force between the beaker and the moving cloth [3–5]. The friction force has a maximum value proportional to the mass m of the beaker and its contents, $F = \mu mg$, where μ is the coefficient of friction (typically a few tenths) and g is the acceleration due to gravity (9.8 m/s^2). According to Newton's second law of motion, this force produces a maximum acceleration of $a = F/m = \mu g$. Thus if you pull the cloth gently (acceleration less than μg), the beaker accelerates along with it, but if you jerk the cloth suddenly (acceleration greater than μg), the cloth is removed before the beaker can accelerate to a significant velocity. What small velocity it does acquire quickly approaches zero because of the friction between the beaker and the table after you remove the cloth. An additional effect that contributes to the success of the demonstration is that the coefficient of sliding friction is less than the coefficient of static friction. Note that the mass of the object cancels in the equation, in contrast to popular misconception [6], and so it is no easier to perform the trick with a heavy object than with a light one. You can illustrate this fact by repeating the demonstration with an empty beaker. If you overdo it with too much mass, you may have trouble pulling hard enough and fast enough on the cloth to produce the required acceleration.

HAZARDS

You should practice this demonstration to develop confidence that the beaker will not break. It is important not to be timid about pulling the cloth and to pull slightly downward. Even so, it is best to set up the demonstration so that if the beaker does break, there is no danger of injury, and to clean up the broken glass in case of an accident. In particular, do not pull the cloth upward or toward the audience.

REFERENCES

1. L. A. Bloomfield, *How Things Work: The Physics of Everyday Life*, John Wiley & Sons: New York (2001).
2. J. L. Ferguson, *Phys. Teach.* **39**, 224 (2001).
3. J. Perez, *Phys. Teach.* **15**, 242 (1977).
4. H. T. Hudson, *Phys. Teach.* **23**, 163 (1985).
5. U. Haber-Schaim and J. H. Dodge, *Phys. Teach.* **29**, 56 (1991).
6. T. L. Liem, *Invitations to Science Inquiry*, Ginn Press: Lexington, MA (1981).

1.7 Pail of Water

A pail of water swung around a vertical circle without the water spilling illustrates Newton's first law of motion.

MATERIALS

- pail of water with secure handle
- short rope (optional)
- bucket of confetti (optional)
- wood block (optional)
- handful of nails (optional)
- tray with several water-filled wine glasses (optional)

PROCEDURE

You can dramatically illustrate the idea of centripetal forces and accelerations as well as Newton's first law of motion (the law of inertia) with a pail partially filled with water [1]. Ask the audience if they think you can turn the pail upside down without spilling the water. Someone will probably say yes, in which case you can threaten to invert it over their head. Swing the pail around a vertical circle without the water spilling, starting about an eighth of a turn in the direction opposite to the rotation.

For a more impressive demonstration, attach a short rope to the bucket handle, but be sure the bucket clears the floor. With a block of wood inside the bucket instead of water, you can turn the bucket upside down without risk of getting wet to illustrate the same principles. A handful of nails in the bucket will make a sound if they leave contact with the bucket near the top of its trajectory. For an amusing aside, surreptitiously exchange buckets with one filled with confetti which you then throw into the audience.

You can also do other more precarious demonstrations. Replace the bucket with a tray containing several water-filled wine glasses. Attach a rope to the tray through holes at each of the four corners and swing the tray around in a circle [1–3]. The main difficulty is in stopping it without breaking the glasses after you have finished swinging it over your head. Let it swing about an eighth of a turn beyond the vertical, and then let it loosely drop to your side. A physics graduate student once volunteered his infant son in a basket but changed his mind when his parental instincts (and perhaps the horror of his wife) edged out his faith in the laws of physics!

DISCUSSION

According to Newton's first law of motion, objects in motion tend to remain in motion with a constant velocity unless acted upon by an external force. In this case, Newton's first law of motion requires the water to continue moving along a tangent to the circle. Thus a force is required to keep it always turning toward the center of the circle. The interpretation of this demonstration is potentially confusing when you consider that at the top of its arc, the water is accelerating downward because of the motion while the force of gravity is also downward. You can explain that $F = ma$ is thus satisfied without the water leaving the bucket. Thus you can use the demonstration also to illustrate Newton's second law of motion. Alternately, explain that the water does fall but that the bucket falls even faster and overtakes it. In the same way, the International Space Station is falling, but it has enough horizontal velocity that the Earth curves away from it just as fast as it is falling. You can use this demonstration to discuss non-inertial (accelerated) frames of reference and inertial (fictitious) forces such as the centrifugal force.

HAZARDS

The hazards of this demonstration are rather obvious. Make sure the area is completely clear of obstructions, that the handle is securely attached to the bucket, and that you have a good grip on it. Do not swing the bucket toward the audience. Also, do not swing it too slowly! Be careful not to hit your leg with the bucket. If water does spill on the floor, remove it so that no one slips on it.

REFERENCES

1. L. A. Bloomfield, *How Things Work: The Physics of Everyday Life*, John Wiley & Sons: New York (2001).

2. T. L. Liem, *Invitations to Science Inquiry*, Ginn Press: Lexington, MA (1981).

3. W. J. Boone and M. K. Roth, *Phys. Teach.* **30**, 348 (1992).

1.8
Revolving Ball and Cut String

A revolving ball released at a certain point in its orbit moves in a straight line in the absence of an external force, illustrating Newton's first law of motion.

MATERIALS

- Styrofoam® ball
- motor with rotating arm
- jackscrew
- razor blade or sharp knife
- light string or thread

PROCEDURE

Attach a Styrofoam® ball, preferably brightly colored or disguised as a baseball, to a light string or thread and spin it around in a circle at the end of an arm driven by a motor. Mount a sharp knife or razor blade on a jackscrew so that you can slowly raise it to the point where it cuts the string. When the string is cut, the ball flies off along a tangent to the circle. You can control the direction that the ball goes by the placement of the knife. Point out that the ball does not keep going in a circle as some people might think. You could have the audience vote beforehand on what will happen.

DISCUSSION

This demonstration illustrates Newton's first law of motion (the law of inertia). When the string is cut, there is no longer a radially inward force causing the ball to move in a circle, and the ball continues along a straight-line trajectory except for the downward effect of gravity.

The threat of cutting the string usually causes audience members to cower in their seats since the ball looks heavy and dangerous and since they intuitively believe that the direction the ball will go is unpredictable. However, the ball will always fly off at right angles to the line between the knife and the center of the circle. Thus you can arrange its trajectory rather precisely. In any event, the ball is light enough so as not to cause any injury.

HAZARDS

The main hazard is to the person doing the demonstration. The jackscrew must be located close to the rotating arm, and you should avoid coming into contact with the rotating arm or having the arm hit the jackscrew. The razor blade on the jackscrew should be located so it will not inadvertently cut you or someone else by brushing against it. Aim the ball so it is not likely to hit someone.

1.9
Inclined Plane

Objects sliding or rolling down an inclined plane illustrate friction and moment of inertia.

MATERIALS

- inclined plane, preferably a meter long or more[1]
- blocks of various materials
- assorted cylinders, hollow spheres, balls, and hoops
- can of soda or soup (optional)
- protractor (optional)
- aluminum channel (optional)
- metronome (optional)

PROCEDURE

Place blocks of various materials on the plane (one at a time or simultaneously), and tip the plane up to an angle at which the blocks just begin to slide. Show that the angle is different for different materials such as smooth wood and rubber. Show that for a given material, the critical angle is independent of the mass of the object and of the area of contact. Show that the angle at which the block starts to slide is slightly greater than the angle required to keep it sliding once it is in motion. You can measure the angle with a protractor. You can also support the block with a string that goes over a pulley attached to a known weight, in which case the friction force is equal to the attached weight.

[1] Available from Frey Scientific, Klinger Educational Products, and Sargent-Welch.

With the plane inclined at a fixed angle, roll cylinders, hollow spheres, balls, and hoops down the plane [1–7]. Before doing this, ask the audience to predict which object will reach the bottom first. Many people will say that the smaller ball will win the race, perhaps reasoning that small animals often outrun larger animals or that the larger ball has more inertia. Others will reason that the heavier ball will win because of the larger gravitational force acting on it. Repeat with objects of different size and the same mass and with objects of the same size and different mass. Show that if the plane is inclined too steeply the objects will slide rather than roll.

Compare the speed of an object rolling without slipping and one sliding without friction (simulated by a large mass with small wheels). Both cases approximately conserve mechanical energy, but the sliding object always reaches the bottom before the rolling object because all the initial potential energy converts into translational energy with none consumed in rotation. You can illustrate that some energy is lost to friction by placing a second identical inclined plane at the bottom of the first so that the objects roll uphill after reaching the bottom and oscillate with ever decreasing amplitude until they come to rest at the seam between the two planes [8].

Roll a can of soda down the plane, and compare its speed to a solid cylinder and to a hoop of similar dimensions. The can of soda should reach the bottom before the solid cylinder but after the hoop since relatively little of the liquid inside the can rotates. Use transparent water bottles filled with various amounts of water and other materials such as dry coffee, oatmeal, rice, or sand. Compare the speed of two soup cans, one containing a watery soup and the other a creamy soup [9]. Depending on the material and the steepness of the incline, the bottle will remain still, roll at constant speed, or accelerate [10, 11]. If you shake the can of soda it will take slightly longer to roll down the plane, presumably because the surface tension of the resulting bubbles increases the transfer of energy to the fluid more efficiently than the fluid itself rubbing against the can [12]. If the incline angle is sufficiently large, you can make a box tumble down it [13].

Repeat the demonstration using hollow spheres and solid balls of various masses and sizes but rolling down the rails of an aluminum channel. In this case the speed depends on the radius of the sphere because the object rolls about an axis that is partway between its center and its surface [9]. Show that two balls, one above the other and placed on the plane touching one another, will roll only when the angle of incline exceeds some critical value that depends on the masses, radii, and coefficients of friction of the balls [14, 15].

You can also use this apparatus to demonstrate that the acceleration of an object rolling down the plane is constant. Put marks on the plane at distances from the starting point in ratios of 1:4:9:16 (the square of the first few integers). Adjust the angle of the incline so that an object released from rest crosses the marks at successive ticks of a metronome. This is the method used by Galileo (1564 – 1642) to demonstrate that objects fall with a constant acceleration.

DISCUSSION

Friction exerts a force in the direction opposite to the direction in which something is moving or induced to move. The friction force is proportional to the normal force, which in this case is the component of the gravitational force on the object in a direction perpendicular to the plane. If the plane is inclined at an angle θ with respect to the horizontal such that the object slides or is just about to slide, the friction force points upward along the plane. It has a magnitude $f = \mu W \cos \theta$, where W (= mg) is the weight and μ is the coefficient of friction. The quantity μ is typically in the range of 0.01 to 1.0 and depends on the materials and the condition (roughness) of the surfaces but not on the area of contact. The coefficient of friction depends somewhat on the speed of the object and, in particular, is greater when the object is at rest (static friction) than when it is in motion (kinetic friction).

The block will begin to slide when the component of the gravitational force in a direction along the plane ($W \sin \theta$) just equals the friction force so that

$$\tan \theta = \sin \theta / \cos \theta = \mu$$

independent of the weight W. A measurement of the critical angle θ at which the block begins to slide thus provides a measure of the coefficient of friction. Friction converts the potential energy of the block at the top of the incline into heat as the block slides down so that it can arrive at the bottom with no potential energy and very little kinetic energy.

When the angle of the plane is sufficiently small that an object can roll down it without slipping, there is no relative velocity between the point on the object that is in contact with the plane and the plane, and thus the friction does no work on the object, and the total mechanical energy is conserved. However, as the object rolls down the plane, its initial potential energy changes into both translational energy of the center of mass and rotational energy. The ratio of the rotational to the translational energy is I/mr^2, where I is the moment of inertia, m is the mass, and r is the radius of the object. The moment of inertia is mr^2 for a hoop, $mr^2/2$ for a solid cylinder, and $2mr^2/5$ for a solid sphere. Thus the hoop acquires the most rotational energy and the least translational energy (and velocity) and thus takes the longest to get down the plane. The sphere is the fastest, and the cylinder is intermediate. Since the initial potential energy and the final kinetic energy are both proportional to mass and independent of the radius, objects of the same shape but different mass and radius move down the plane at the same rate. If the plane is inclined too steeply, the object will slip, friction will do work, and the rate at which the objects roll is more difficult to predict. The viscosity of a liquid inside the rolling object tends to dissipate energy and slows the translational motion.

HAZARDS

There are no hazards with this demonstration except to ensure that when the objects reach the bottom of the incline or if they fall off the edge, something catches them in a manner that prevents them from doing any damage.

REFERENCES

1. W. P. Berggren and M. E. Gardner, *Am. J. Phys.* **9**, 243 (1941).
2. W. B. Pietenpol, *Am. J. Phys.* **13**, 260 (1945).
3. J. S. Miller, *Physics Fun and Demonstrations*, Central Scientific Company: Chicago (1974).
4. M. Salete, S. C. P. Leite, and C. A. N. Conde, *Phys. Educ.* **9**, 426 (1974).
5. R. H. March, *Phys. Teach.* **26**, 297 (1988).
6. G. W. Ficken, *Phys. Teach.* **32**, 197 (1994).
7. R. P. McCall, *Phys. Teach.* **42**, 212 (2004).
8. C. Melvin, *Phys. Teach.* **40**, 222 (2002).
9. C. R. Stannarad, T. P. O'Brien, and A. J. Telesca, *Phys. Teach.* **30**, 526 (1992).
10. D. Kagan, *Phys. Teach.* **39**, 290 (2001).
11. G. D. Nicas, *Am. J. Phys.* **57**, 907 (1998).
12. A. Cromer, *Phys. Teach.* **34**, 48 (1996).
13. A. M. Nunes and J. P. Silva, *Am. J. Phys.* **68**, 1042 (2000).
14. W. Dindorf, *Phys. Teach.* **37**, 184 (1999).
15. A. J. Mallinckrodt, *Phys. Teach.* **37**, 463 (1999).

1.10
Bowling Ball Pendulum

A bowling ball suspended from the ceiling by a thin stainless steel wire illustrates the simple harmonic oscillator and the conservation of energy.

MATERIALS

- bowling ball: 16 pounds, 8.5-inch diameter
- thin, stainless steel wire
- footstool or stepladder
- table (optional)
- stopwatch or metronome (optional)
- softball (optional)

PROCEDURE

A volunteer, solicited from the audience, stands with the back of his head against a wall (or a table turned on its end if a wall is not conveniently located) and a bowling ball suspended from a wire from the ceiling held snugly against his nose, perhaps standing on a footstool or stepladder for increased height. The bowling ball is released (not pushed!), and you ask the volunteer to put his hands by his side and not to move while the ball swings back to within a few centimeters of his nose. It is best if the pendulum is as long as possible, if the volunteer is far back from the point of suspension, and if the audience views from the side. Mark a point on the wall to indicate the proper position for his head, and choose a volunteer of approximately the correct height so that his head is in the proper position. Glasses on the volunteer add to the drama. Glassware or other fragile apparatus that happens to be just centimeters below the trajectory of the ball heightens the excitement.

As an amusing variant, give the ball a push when released, and have the subject move aside at the last moment as the ball crashes into the wall. This variant is not recommended for an unsuspecting volunteer, and you might prefer to do it yourself or with a trusted assistant. While waiting for the ball to return, recite, "I believe in the conservation of energy…"

With a stopwatch, or simply by counting the seconds or the ticks of a metronome, you can determine the time it takes for the ball to swing out and back. You can then repeat the demonstration with balls of different sizes and masses (a softball works well) to show that the period, but not necessarily the frictional losses, depend only on the length of the pendulum. Ask the audience to vote on whether a lighter ball will take a longer or a shorter time to return to its starting point. Most people will say a shorter time, but the time is essentially the same. Show that the period of the pendulum does depend on its length.

By choosing different starting points for the ball, or by observing the period carefully as the ball slowly comes to rest, you can demonstrate that the period does not depend on the amplitude of the motion, provided the amplitude is small. You can use the demonstration to emphasize the deterministic nature of classical physics and the accuracy with which one can predict certain (but not all) physical systems. Such a regular, predictable motion contrasts with chaotic motion, which is deterministic but unpredictable because of the sensitive dependence on initial conditions [1].

A bowling ball suspended from above can also serve as a good demonstration of Newton's first law of motion (the law of inertia) [2]. It takes a large horizontal force to accelerate the ball. On the other hand, a softball hanging beside the bowling ball requires a much smaller force to accelerate it. Thus the mass of the ball is a measure of its inertia.

This might also be a good time to emphasize that the laws of physics, which guarantee that you can do this demonstration safely, are different from the kind of laws that the police enforce. Natural laws, such as Newton's laws and the conservation of energy, are universal and eternal. They have been the same for over ten billion (10^{10}) years, and will probably remain true for at least that long into the future if not forever. Not only are they the same in every country but on every planet and in every galaxy, unlike human laws (do not say the "laws of man," which sounds sexist) that differ from place to place and time to time. You might use this thought as motivation for students to study physics. The law of energy conservation is perhaps the most important law in physics and maybe even in all of science.

DISCUSSION

This demonstration illustrates the transformation of potential energy when you lift the ball above its resting position into kinetic energy as the ball acquires velocity, and back, with mechanical energy losses due to friction with the air, which converts the mechanical energy into heat energy. A sixteen-pound bowling ball has about 70 joules of potential energy for each meter of height. Ordinary bowling balls weigh between ten and sixteen pounds.

By measuring or estimating the amount by which the ball misses the volunteer's nose, you can determine the fractional energy loss per cycle [3]. The period of the pendulum is $2\pi\sqrt{L/g}$, where L is the length of the pendulum and g is the acceleration due to gravity (9.8 m/s^2). The period does not depend on the mass of the pendulum or on the amplitude of the motion, at least as long as the amplitude is sufficiently small. If the pendulum swings to an angle of 45° from the vertical, the period is about 4% greater than calculated, and at 90°, the period is about 18% greater. The large-angle formula for the period involves an elliptic integral [4, 5], but good approximate formulas exist [6, 7].

One reason there is a particular rate at which it is comfortable to walk is that your legs are pendulums, and it is easiest to walk at a rate such that they swing at their natural resonant frequency. The arms swing at the same frequency because they are nearly the same length as the legs. When you run, you instinctively bend at your elbows so that the arms swing at a higher frequency corresponding to the rate at which your legs are moving.

HAZARDS

The wire should have a tensile strength at least several times the weight of the ball and should be anchored securely. Keep the wire taut at all times. Make sure the ball cannot strike anything during its swing, and stand in a position so that you can stop it if the volunteer gives it a push rather than just releases it. This demonstration is potentially dangerous if not done properly since the collision of the bowling ball with one's face will cause serious injury. Do not allow anyone to try it without proper supervision.

REFERENCES

1. J. C. Sprott, *Chaos and Time-Series Analysis*, Oxford University Press: Oxford (2003).
2. G. Hodgson, *Phys. Teach.* **32**, 117 (1994).
3. J. H. Head, *Phys. Teach.* **33**, 10 (1995).
4. J. B. Marion, *Classical Dynamics of Particles and Systems*, Harcourt Brace Jovanovich: San Diego (1970).
5. M. L. Boas, *Mathematical Methods in the Physical Science*, John Wiley & Sons: New York (1966).
6. T. F. Zheng, M. Mears, D. Hall, D. Pushkin, S. Jorgensen, K. R. Banks, R. J. Hyatt, and Y. M. Shleyzer, *Phys. Teach.* **32**, 248 (1994).
7. R. B. Kidd and S. L. Fogg, *Phys. Teach.* **40**, 81 (2002).

1.11 Come-back Can

A can, when rolled across a table, comes to rest and then rolls back to where it started, illustrating the concept of stored internal energy.

MATERIALS

- cylindrical can with removable, opaque lid
- rubber band
- weight with hole though its center

PROCEDURE

Construct a can with a rubber band strung between the center of its ends and a weight connected to the band so that the band winds up as the can rolls [1–3]. The can comes to rest and then rolls back to where it started. The appearance is that the table is not level, but you can roll it in either direction with the same result. It helps to rotate the can a turn or two before releasing it to compensate for the frictional losses as it rolls. This also allows the can to roll up a slight incline. One end of the can should be easily removable to reveal its contents and to illustrate its operation. In a variation of the demonstration, rotate the can gently, and then set it on its end to make it "dance."

DISCUSSION

This demonstration illustrates the conversion of kinetic to potential energy and back. The potential energy is stored internally in the twisted rubber band. On the microscopic level, the energy is stored in the electric potential of the charged particles that make up the molecules of the rubber. You can make analogies to winding a watch, to filling an automobile gasoline tank, to the stored energy in atoms and molecules, and to the energy of mass itself ($E = mc^2$).

From the standpoint of the special theory of relativity, the rest mass of the can and its internal mechanism increases slightly as the rubber band winds up, and it is this increased mass that is converted into kinetic energy when the can begins to roll from a stop [4]. You can estimate the increase in mass Δm from

$$\Delta mc^2 = mv^2/2$$

to show why the mass increase is not normally detectable for objects moving slowly compared with the speed of light. For example, a can with an initial speed of 1 m/s increases its mass by less than 1 part in 10^{17}! Most people are familiar with the process of converting mass into other forms of energy in nuclear fission and fusion, but the reverse can also happen, as it does here. A more dramatic example is electron-positron pair production when a photon of sufficient energy interacts with matter and produces new particles. As Albert Einstein (1879 – 1955) asserted, "the inert mass of a closed system is identical to its energy" [5].

HAZARDS

There are no significant hazards with this demonstration.

REFERENCES

1. T. L. Liem, *Invitations to Science Inquiry*, Ginn Press: Lexington, MA (1981).
2. J. P. VanCleave, *Teaching the Fun of Physics*, Prentice-Hall: New York (1985).
3. B. Doerrie, J. Doerrie, H. Cook, D. Wilson, N. Horton, and S. Christian, *Phys. Teach.* **33**, 388 (1995).
4. E. Hecht, *Phys. Teach.* **41**, 486 (2003).
5. A. Einstein, *Philosopher-Scientist*, edited by Paul Schilpp, Harper & Row: New York (1959).

1.12 Collision Balls

Five stainless steel balls suspended in a row from above demonstrate conservation of momentum and energy in nearly elastic collisions.

MATERIALS

- collision balls[1]
- sheet of paper (optional)

PROCEDURE

Most people have seen the toy consisting of five (or more) identical stainless steel balls suspended from above and arranged in a row with each ball just touching its neighbor(s). When you pull one ball back and release it, it collides with the row of balls, ejecting the one at the far end. After demonstrating this behavior, ask what would happen if you pull back and release two balls. Most people will say that two balls will be ejected from the other end. Then ask what will happen if three balls are used. It is not quite so obvious that three balls will be ejected since there are only two balls left, but that is exactly what happens.

Explain that this demonstration illustrates the conservation of momentum mv. However, there are many ways to conserve momentum besides the one shown here. For example, one could have two balls ejected with half the velocity. The peculiar behavior results from the fact that the collisions are nearly elastic, and thus kinetic energy $mv^2/2$ is also approximately conserved in the collisions. The only way to conserve both quantities simultaneously is to have the mass and velocity individually conserved. The kinetic energy is converted into potential energy as the ball swings outward and comes to rest, after which it swings back and starts the process over but in the opposite direction. You must accurately align the balls for this demonstration to work properly.

Ask the audience how they could prove that the collisions are not perfectly elastic. Give a hint that they can do so with their eyes closed. Some people may realize that the sound of the collision represents a conversion of mechanical energy into sound energy. The energy loss through sound is very small, and in fact accounts for only a tiny fraction of the total energy loss, most of which consists of heating the balls. Of course the sound energy also eventually changes into heat once the echoes die down. With sufficiently

---

[1] Available from Carolina Biological Supply Company, Frey Scientific, Sargent-Welch, and Ward's Science.

large balls moving with sufficient speed, you should be able to burn a hole in a sheet of paper placed between them, illustrating the heat produced by the collision.

You can also demonstrate collisions between balls of unequal mass with this or another apparatus. Try placing wax on four of the balls and sticking them together so that they behave like a single mass four times greater than the mass of the ball with which they collide. Alternately, place wax on all the balls to illustrate perfectly inelastic collisions. For an impressive version of the demonstration, use bowling balls [1].

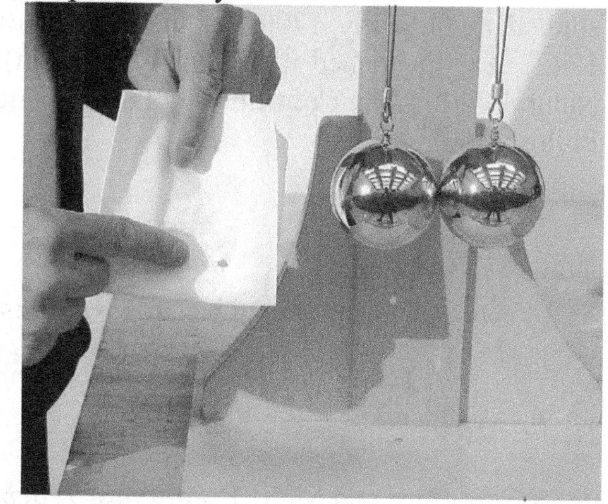

Collisions balls are sometimes called "momentum balls" or "Newton's Cradle," and they do demonstrate Newton's laws, but you should avoid calling them "Newton's balls" except perhaps for the amusement of a mature audience.

DISCUSSION

Consider first the head-on (one-dimensional) collision between two identical balls of mass m, one moving with initial velocity v_{1i} and the other at rest with $v_{2i} = 0$. After the collision, they are moving with final velocity v_{1f} and v_{2f}, respectively. Conservation of momentum requires $v_{1f} + v_{2f} = v_{1i}$, and conservation of energy requires $v_{1f}^2 + v_{2f}^2 = v_{1i}^2$. Squaring the first equation and then subtracting the second from it gives $v_{1f}v_{2f} = 0$. Thus momentum and energy can be conserved only if $v_{1f} = 0$ or $v_{2f} = 0$ or both. The first case is the one demonstrated here. The second case would correspond to the balls missing one another entirely (no collision), and the third case would correspond to the balls sitting at rest.

You can think of the demonstration as a succession of four such collisions. It may help to think of the balls as having a small initial gap between them [2]. With each collision, according to Newton's third law of motion (action and reaction), there is a decelerating force on the moving ball exerted by the ball with which it collides that just brings it to rest until the last ball is reached, in which case there is no ball to stop it. This model is obviously at odds with reality, but the detailed analysis requires a consideration of the elasticity of the balls that approximately follows the force law $F = -kx^{3/2}$ rather than $F = -kx$ (Hooke's law) [3].

To see that as many balls are ejected as are launched, let m_1 be the total mass of the balls launched with velocity v_{1i} and m_2 be the total mass of the balls ejected with velocity v_{2f}. To find the condition for which $v_{1f} = 0$, use conservation of momentum $m_2v_{2f} = m_1v_{1i}$ and conservation of energy $m_2v_{2f}^2 = m_1v_{1i}^2$. Square the first equation and subtract m_1 times the second equation to see that $m_2 = m_1$ as demonstrated.

If the balls are treated as a unit in various combinations rather than separately, there are actually other solutions that satisfy both momentum and energy conservation. Even given the initial conditions of all five balls, it is not possible to calculate the final velocity of all the balls, since five equations are required and the conservation laws only provide two equations. The correct explanation requires a detailed examination of the force that each ball exerts on its neighbor, and it is rather involved [4–11]. This apparatus was studied in the early 1700s by the Dutch lawyer Willem Jacob 's Gravesande (1688 – 1742), who published it (originally in Latin) under the title "Mathematical Elements of Natural Philosophy Confirm'd by Experiments; or, An Introduction to Sir Isaac Newton's Philosophy."

HAZARDS

There are no significant hazards with this demonstration.

REFERENCES

1. W. J. Boone and M. K. Roth, *Phys. Teach.* **30**, 348 (1992).
2. S. Chapman, *Am. J. Phys.* **9**, 357 (1941).
3. B. Leroy, *Am. J. Phys.* **53**, 346 (1985).
4. S. Chapman, *Am. J. Phys.* **28**, 705 (1960).
5. L. Flansburg and K. Hudnut, *Am. J. Phys.* **47**, 911 (1979).
6. F. Herrmann and M. Schmälzle, *Am. J. Phys.* **49**, 761 (1981).
7. F. Herrmann and M. Seitz, *Am. J. Phys.* **50**, 977 (1982).
8. R. Ehrlich, *Phys. Teach.* **34**, 181 (1996).
9. J. D. Gavenda and J. R. Edington, *Phys. Teach.* **35**, 411 (1997).
10. C. T. Tindle, *Phys. Teach.* **36**, 344 (1998).
11. D. Kagan and C. Gaffney, *Phys. Teach.* **40**, 280 (2002).

1.13 Rockets

Toy rockets illustrate Newton's third law of motion and the conservation of momentum.

MATERIALS

- toy rockets[1]
- safety glasses
- S-shaped lawn sprinkler and compressed air (optional)
- 5-gallon plastic water jug with 35 ml of ethyl alcohol (optional)
- carbon dioxide fire extinguisher and rotating platform (optional)
- basketball (optional)
- balloon (optional)
- radiometer[2] with bright light, match, or candle (optional)

PROCEDURE

Toy rockets are available from many toy stores as well as from vendors of scientific apparatus. Rockets suitable for indoor demonstrations use compressed air and water [1–4] or a cartridge of compressed carbon dioxide as the propellant. Solid propellant rockets are most safely demonstrated outdoors [5, 6]. You can perform more modest demonstrations with balloons that rapidly expel their air, although the trajectory is much more complicated (in fact chaotic) in such cases. With a water rocket, you can demonstrate the effect of using different proportions of air and water. The rocket will work, but not very well with only air. For added interest, have someone try to catch the

[1] Available from Arbor Scientific, Carolina Biological Supply Company, Fisher Science Education, Frey Scientific, Nasco, Science First, Estes-Cox Corporation, PASCO Scientific, Sargent-Welch, Ward's Science, and Scientifics.

[2] Available from American 3B Scientific, Arbor Scientific, Carolina Biological Supply Company, Educational Innovations, Frey Scientific, Sargent-Welch, Ward's Science, and Scientifics.

rocket in a fish net. You can attach the rocket to a wire strung across the room or up to the ceiling to control the direction that it goes. You can also place the rocket on the end of a rotating arm to control its trajectory [1].

An S-shaped lawn sprinkler run on compressed air provides an alternate demonstration of the same principle as well as a demonstration of the conservation of angular momentum. A crowd thinking the sprinkler is connected to water rather than air quickly comes to attention when you turn it on!

Ask the audience what would happen if the flow of water or air were into the nozzles of the lawn sprinkler rather than out. You can do this demonstration by connecting the lawn sprinkler to a vacuum pump or to a water pump with the sprinkler submerged under water in an aquarium [7]. Typically there is little or no rotation because the air or water enters the nozzles from all directions rather than in a jet, but the sprinkler will tend to turn in the opposite direction, but slowly. Tell the story of how the famous physicist Richard Feynman (1918 – 1988) used to pose this question to his colleagues and flooded the cyclotron laboratory at Princeton University late one night doing the experiment, being forever banned from the laboratory [8].

You can make quite an impressive rocket with a 5-gallon plastic water jug in which you place about 35 ml of methyl or ethyl alcohol. Place it on a horizontal surface and bring a lighted match near its opening, while standing to the side. It should go twenty feet or more with a loud noise and plume that is especially impressive in subdued illumination. You can launch it off the edge of a table. Experiment with different amounts of alcohol until you find a comfortable amount, but it is important to purge the bottle completely of residual fumes between each launching, for example, by blowing compressed air into the bottle. Swish the alcohol around after placing it in the jug to coat the walls of the jug with liquid and wait a few minutes with the bottle capped to allow some of the alcohol to vaporize. For a more modest demonstration, use a one-liter plastic soft drink bottle.

A carbon dioxide fire extinguisher provides enough thrust for some impressive rocket demonstrations. While sitting in a swivel chair or standing on a rotating platform [9–11], you can be made to rotate rapidly. You can construct a rotating platform using a junk automobile water pump [12].

A carbon dioxide fire extinguisher mounted on a cart [13, 14] or bicycle will propel you for several hundred feet. Roller skates are another possibility. You can have a special guest (perhaps Isaac Newton or Albert Einstein) enter or exit on such a device, perhaps accompanied by a discussion of time dilation for objects moving at close to the speed of light.

You can effectively introduce the discussion by contrasting the rocket with all other vehicles that move by exerting a force on the medium across which or through which they move. An automobile moves forward by virtue of the reaction force between its wheels and the ground. A boat moves forward by the reaction force on the propeller or paddle. An airplane moves forward by the reaction force of the air that it pushes backward by the propeller or turbine blades. Contrast a rocket engine, which carries its own exhaust, with a jet engine, which exhausts air sucked in through its intake.

Many people misunderstand and think that a rocket works by pushing against something. In fact, you can think of it as working by pushing against its own exhaust. You can illustrate the operation by throwing a basketball while standing on a platform that is free to rotate, in which case you are pushing against the basketball, but the basketball eventually has to leave contact with your body. You could reverse the demonstration, and have someone toss the basketball to you while you are on the platform, but they should not throw it directly at you. This variant can be a little hazardous if you or the volunteer fail to catch the basketball. A little "Sweet Georgia Brown" music fits in well here.

A less familiar example of a rocket is a radiometer (also called "Crookes radiometer"), which consists of a number of small vanes colored black on one side and white or silver on the other and mounted inside a partially evacuated glass sphere. The vanes rotate when a bright light shines on them because the black surface becomes hotter than the white surface, imparting more energy to the air molecules that collide with it and bounce off. The rotation is not due to the radiation pressure of the light, since that would cause the vanes to rotate in the opposite direction because the light reflects from the white surface and hence imparts more momentum to it than to the black surface that absorbs most of the light. Mention that sunlight exerts a total radiation pressure of about 4.5 pounds on the Earth.

You can also make the radiometer vanes rotate using a nearby match or candle. The radiometer illustrates the conversion of light energy into heat energy and then into the energy of motion (kinetic energy), and finally back to heat again through the friction of the moving vanes with the air.

DISCUSSION

The rocket moves forward by the reaction force of the ejected exhaust of the rocket itself (Newton's third law of motion). Stated differently, the total momentum of the rocket and exhaust is constant. The general situation is described by the (nonrelativistic) rocket equation

$$v_f = v_i + v_e \ln(m_i/m_f)$$

where v_i and v_f are the initial and final velocities of the rocket, respectively, v_e is the velocity of the exhaust relative to the rocket, and m_i and m_f are the initial and final masses of the rocket, respectively, considering the loss of mass due to the ejected exhaust. Alternately, you can explain the propulsion of the rocket in terms of the center of mass of the rocket and exhaust moving at a constant velocity. The process can be described as a perfectly inelastic collision but in reverse (so that the masses stick together before the collision rather than after), or as a heat engine in which the heat of the burning fuel is converted into kinetic energy of the rocket. The instantaneous thrust of the rocket is the magnitude of the quantity $v_e dm/dt$, which has the same units as force and is in the opposite direction from the force on the exhaust. The thrust must exceed the weight (mg) to get off the ground if a rocket fired vertically upward. The kinetic energy of the rocket comes from the potential energy stored in the form of a compressed gas (the product of pressure and volume) or chemical bonds, depending upon the type of propellant.

When rockets were first proposed as a means of space travel by Robert Goddard (1882 – 1945), the *New York Times* (January 13, 1920) printed a rebuttal, claiming that such rockets would never work in the vacuum of outer space because there was no medium against which to push. Forty-nine years later (July 17, 1969), while the *Apollo 11* astronauts were en route to the Moon, the *Times* published a retraction, expressing regret for its error.

HAZARDS

Toy rockets can be dangerous if aimed at someone. You should know the characteristics of the rocket that you will use and practice firing it in a direction that will not hurt anyone or damage nearby equipment. Wear safety glasses when you are working with rockets. If you toss a basketball to someone or have someone toss it to you, be sure it is thrown in a direction where it will not cause harm if it is not caught. It is not recommended to use a volunteer on the rotating platform, but if you do, locate the platform far from any objects that could cause injury if the person falls, and stand nearby to assist if the volunteer loses his or her balance. If you use a fire extinguisher on the rotating platform, hold it close to your body to reduce the torque. If you use a tricycle, be sure to practice, especially making sharp turns, since your instincts from riding a bicycle will make you lean, causing the tricycle to topple over.

REFERENCES

1. J. S. Miller, *Physics Fun and Demonstrations*, Central Scientific Company: Chicago (1974).
2. W. Esler and D. Sanford, *Sci. Teach.* **56**, 20 (May 1989).
3. W. P. Palmer, *Aust. Sci. Teach. J.* **37**, 34 (1991).
4. D. Kagan, L. Buchholtz, and L. Klein, *Phys. Teach.* **33**, 150 (1995).
5. R. L. Cannon, *Model Rocketry, The Space Age Teaching Aid: A Teacher's Guide* (3rd ed.), Estes Industries: Penrose, CO (1990).
6. S. A. Widmark, *Phys. Teach.* **36**, 148 (1998).
7. M. R. Collier, R. A. Ferrell, and R. E. Berg, *Am. J. Phys.* **59**, 349 (1991).
8. R. P. Feynman, *Surely You're Joking, Mr. Feynman!*, Bantam: Toronto (1986), pp. 51–53.
9. G. D. Beadle, *Phys. Teach.* **27**, 488 (1989).
10. N. R. Greene, *Phys. Teach.* **35**, 431 (1997).
11. A. Bryant, *Phys. Teach.* **38**, 476 (2000).
12. N. R. Greene, *Phys. Teach.* **35**, 431 (1997).
13. E. Jones, *Phys. Teach.* **14**, 112 (1976).
14. F. Holt and G. Amann, *Phys. Teach.* **27**, 560 (1989).

1.14 Rolling Chain

A rotating chain retains its circular shape as it rolls across the lecture bench and objects in its path.

MATERIALS

- small, flexible chain (about 60 cm long) wrapped into a loop
- wooden disk to fit the chain
- electric motor
- wooden stick
- inclined ramp
- safety glasses
- pan of water (optional)

PROCEDURE

Wrap a chain around a wooden cylinder and spin it up to a large rotational velocity with an electric motor, and then nudge it off the cylinder with a stick [1, 2]. The chain will retain its circular shape and will roll across the lecture bench and across objects in its path until eventually coming to rest in a pile on the floor some distance away. You can make the chain climb a ramp and fly across the room. A 20-cm-diameter cylinder rotating at 2,500 rpm makes an effective demonstration.

You can point out that the chain takes some time to begin rotating with the cylinder because it must be accelerated by the friction force between it and the cylinder. Similarly, when the chain comes off the cylinder, it slips momentarily on the surface beneath it before the friction force accelerates it forward.

You can also lower the spinning cylinder into a pan of water with the chain removed to illustrate that when the water breaks free from the cylinder, it flies off on a tangent as one would expect from Newton's first law of motion (the law of inertia) [3].

DISCUSSION

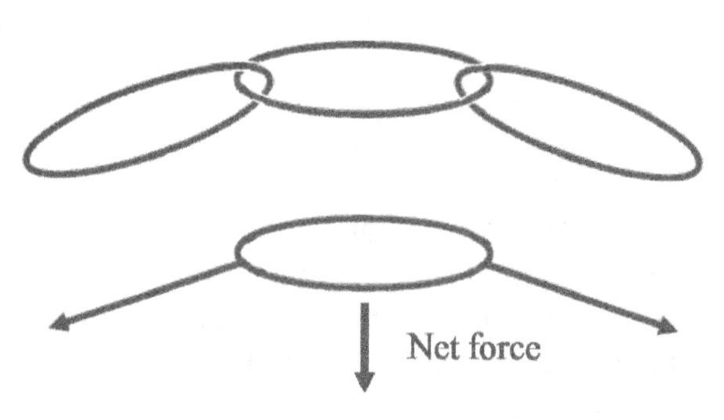

The chain retains its shape because of the inertia of each of its links, which tend to move in a straight line that is tangent to the circle. This tendency is often ascribed to an outward centrifugal force, but it is not a force at all. In fact the force is centripetal since it acts toward the center of the circle in order to keep the chain from flying apart. You can understand the force in terms of the tension in the chain, such that the two adjacent links pull on each link. Since the chain is curved, these forces have a component toward the center of the circle. Alternately, you can describe the behavior of the chain in terms of the conservation of angular momentum, which requires it to continue rotating until sufficient frictional torque brings it to rest.

HAZARDS

A reasonably light chain rotating at a modest velocity is relatively harmless and can even be amusing if it tries to run over someone, but even so it is best to aim it toward a wall and away from fragile equipment. It will travel in a reasonably straight line unless deflected. Grasp the stick securely so that it is not ejected into the audience when it comes into contact with the chain and spool. It is possible for the stick to splinter or the chain to break or fly off in the wrong direction, and so you should wear safety glasses.

REFERENCES

1. H. A. Robinson, Ed., *Lecture Demonstrations in Physics*, American Institute of Physics: New York (1963).

2. J. S. Miller, *Physics Fun and Demonstrations*, Central Scientific Company: Chicago (1974).

3. D. F. Collins, *Phys. Teach.* **42**, 79 (2004).

1.15 Moving Spool

A large wooden spool (or yo-yo) with a string wound around it from below can be made to move either in the direction in which the string is pulled or in the opposite direction depending on the angle of the string.

MATERIALS

- wooden spool or yo-yo
- string
- calipers and protractor (optional)

PROCEDURE

You can effectively introduce this demonstration [1–6] by asking the audience whether they have ever had a yo-yo and showing them one. Point out that your spool is like a yo-yo, and ask them to predict whether it will move forward or backward when you place in on the table and pull the string. Hold the string at an angle where the result is ambiguous when asking the question. Whichever way the majority of the audience votes, you can make the spool go in the opposite direction by pulling the string at the appropriate angle. A small angle θ between the string and the horizontal will make the spool move in the direction of the pull, and a large θ will make the spool move away from the pull. A change in angle so small that the audience does not notice will reverse the direction. The behavior of the spool is quite mysterious, but point out that this is not a magic trick but rather a principle of physics. Show that there is a critical angle at which the spool slips without rolling in either direction. Demonstrate what happens when you try to pull the spool in the opposite direction (an angle θ greater than 90°).

You should use a spool whose runners are far apart and have considerable friction. Wooden spools are usually better than metal ones because they have more friction. You should center the string on the spool so that the spool moves in a straight line. Electrical wire spools are available in a variety of sizes from many sources.

DISCUSSION

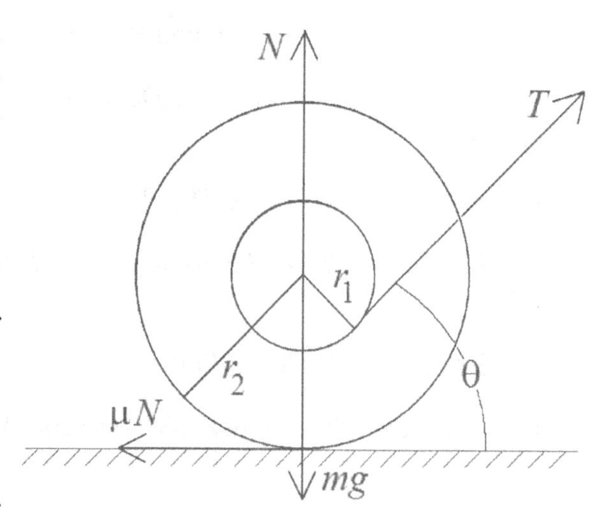

The explanation involves Newton's second law of motion ($F = ma$) and a consideration of the forces and torques on the spool (see the diagram). It is easiest to consider the case where you pull the string with a force and at an angle such that the spool is just on the verge of slipping without rolling. There are four forces: the weight (mg), the upward normal force of the table (N), the tension in the string (T), and the friction force (μN). If the spool is not yet moving, the net horizontal force is zero, or $T \cos \theta = \mu N$. Only two of the forces produce a torque about the center of the spool (T and μN), and these torques must be equal and opposite if the spool is to slip rather than rotate. Equating the torques gives $r_1 T = r_2 \mu N$. Dividing this equation into the previous one gives $\cos \theta = r_1/r_2$. Thus the critical angle that determines which way the spool will rotate depends only on the ratio of the two radii and is independent of the mass of the spool, the tension in the string, and the coefficient of friction. With calipers and a protractor, you can verify the predicted critical angle. This is an especially interesting example of the equilibrium of a rigid body.

An alternate and simpler way to understand the behavior of the spool is to consider the torque about the point where the spool touches the table. The only force that contributes to this torque is the tension in the string, and its direction is either clockwise or counterclockwise depending on the direction of the moment arm. A simple diagram will show the angle at which the moment arm is zero.

A curious feature of this demonstration is that for angles less than the value calculated above, the friction force reverses direction and points in the same direction as the string, actually helping to accelerate the spool [7]. The same thing happens for the tires of an automobile, where the friction force pushes the car forward by preventing the wheels from slipping. Just as there is an angle that allows the spool to slip without rolling, there is a second larger angle in the backward direction at which it never slips (the friction force is zero), provided you do not pull so hard that the spool loses contact with the table [8]. This angle is given by $\cos \theta = -mr_1^3/Ir_2$ where I is the moment of inertia of the spool about its axis.

HAZARDS

There are no significant hazards with this demonstration.

REFERENCES

1. G. M. Martin, *Am. J. Phys.* **26**, 194 (1958).

2. W. A. Hilton, *Phys. Teach.* **2**, 139 (1964).

3. T. J. Parmley, J. I. Swigart, and R. L. Doran, *Phys. Teach.* **4**, 76 (1966).

4. J. S. Miller, *Physics Fun and Demonstrations*, Central Scientific Company: Chicago (1974).

5. P. Chagnon, *Phys. Teach.* **31**, 32 (1993).

6. R. Ehrlich, *Why Toast Lands Jelly-Side Down*, Princeton University Press: Princeton, NJ (1997).

7. A. Pinto and M. Fiolhais, *Physics Education* **36**, 250 (2001).

8. C. E. Mungan, *Phys. Teach.* **39**, 481 (2001).

1.16
Bicycle Wheel Gyroscope

A bicycle wheel attached to a wire with a fishing line swivel and suspended from a support makes an impressive gyroscope.

MATERIALS

- bicycle wheel gyroscope[1]
- stainless steel wire
- fishing line swivel
- rotating stool or platform[2] or swivel chair (optional)
- hand-held barbells (optional)
- suitcase with internal gyroscope (optional)

PROCEDURE

Take out a bicycle wheel and ask who has one of these at home. All the children will immediately recognize it and will wave their hands. Point out that it is a bicycle wheel, but it is a special one because it contains concrete. Spin the bicycle wheel up to high speed by hand or with a rope, and suspend it from the ceiling by a wire with a fishing line swivel attached to one end of its axle [1–3]. The concrete gives the wheel more angular momentum and makes the demonstration more dramatic. The fishing line swivel allows it to precess without twisting the supporting wire.

As the gyroscope precesses, you can make its axis bob up and down in a motion known as nutation. If a bicycle wheel gyroscope is not available, you can show the same effects on a smaller scale with a toy gyroscope or even a toy top [4].

Ask a strong volunteer from the audience to hold the spinning bicycle wheel by one end of its axle horizontally at arms' length and then raise it vertically over his head, first with the wheel not spinning and then with it rotating rapidly. It is difficult to do this

---

[1] Available from American 3B Scientific, Arbor Scientific, Carolina Biological Supply Company, Fisher Science Education, Frey Scientific, Klinger Educational Products, PASCO Scientific, Sargent-Welch, and Science First.

[2] Available from American 3B Scientific, Arbor Scientific, Carolina Biological Supply Company, Fisher Science Education, Frey Scientific, PASCO Scientific, Sargent-Welch, and Science First.

without a bit of practice. Mount a spinning gyroscope inside a suitcase to provide a particularly spectacular and unforgettable demonstration to the person who tries to turn abruptly while carrying the suitcase.

Hold the bicycle wheel in your hands while sitting on a rotating stool or swivel chair or while standing on a rotating platform [5–7] to illustrate Newton's third law of motion (action and reaction) and the conservation of angular momentum. You can make the stool rotate one direction or the other by turning the axis of the bicycle wheel in different directions. While standing on the stool, hand the wheel to someone standing on the floor who inverts it and hands it back to you. You then invert it again, and hand it back to the person on the floor who inverts it again and so forth until the stool spins quite rapidly. The bicycle wheel is imparting quanta of angular momentum to you. Merry-go-round music works well here.

The rotating platform can serve to demonstrate the conservation of angular momentum using a pair of hand-held barbells [8]. Start with the barbells at arms' length and have someone give you a gentle spin. Then pull the barbells in toward your body to make yourself spin faster. You can slow down again by extending your arms. This effect is familiar to anyone who has observed a figure skater doing spins on the ice. As the moment of inertia changes, the angular velocity must also change so as to keep their product (the angular momentum) constant. You can also demonstrate the Coriolis force by showing that as you relax your arms and let them fall toward your body while rotating on the platform, one arm goes forward and the other backward [9].

The gyroscope has many interesting properties. You can show that its axis will remain horizontal so long as you allow it to precess. When you stop the precession, it falls. You must apply the force required to make it move in a particular direction at right angles to that direction. When swung like a pendulum, the wheel tends to remain in a plane. This is the principle of inertial guidance of rockets, the gyrocompass, and other navigational instruments [10]. It is also the reason a spinning football is more stable and easier to catch.

You can point out that the gyroscopic action of the wheels is one (but not the main) reason a bicycle remains upright. In that case there is no precession since the wheel is suspended from its center of gravity. You can roll the bicycle wheel across the floor to illustrate that it stays upright much longer than it would if released from rest. The static and dynamic stability of bicycles makes an interesting digression [11–15].

You can build a larger version of the gyroscope using a water-filled automobile tire and other inexpensive parts easily obtained from junkyards [16].

DISCUSSION

The gyroscope provides an interesting and unusual example of the conservation of angular momentum. The angular momentum is a vector pointing along the axis about which the gyroscope spins (in a sense given by the right-hand rule). In the absence of external torques, the direction as well as the magnitude of this vector will remain constant. Friction produces a torque that decreases the magnitude of the vector and eventually causes the gyroscope to stop spinning. Gravity produces a torque perpendicular to both the axis of the gyroscope and the vertical, and thus causes the horizontal precession. On a less abstract level, you can explain the precession in terms of the downward pull of gravity that tries to make the wheel rotate faster at the bottom than at the top. Since the wheel is rigid, this can happen only if the wheel moves horizontally in the direction in which the bottom of the wheel is spinning. The Earth is a large gyroscope that precesses once every 26,000 years due to the gravitational torque exerted by the Sun on the slight bulge at the equator. This precession may be at least partially responsible for the onset of the Ice Ages.

Note that the frequency of precession is inversely proportional to the frequency at which the gyroscope is spinning. You can illustrate this fact by observing carefully the precession as the gyroscope slows down. Furthermore, the precession frequency is independent of the angle that the axis makes with the horizontal. The torque is greatest when the axis is horizontal, but so also is the distance it has to move to precess once around, and the effects just cancel.

The kinetic energy associated with the precession has to come from somewhere. It comes from the gravitational potential energy of the gyroscope itself. When you release the gyroscope from an initial fixed horizontal position, it starts to fall in the usual manner. This falling motion rapidly transforms into precession, with the center of mass slightly lower than it was initially. As it falls, it overshoots its equilibrium position slightly and oscillates up and down about this equilibrium, resulting in nutation. The nutation usually damps out rather quickly, but you can excite it by a rapid upward or downward impulsive force on the free end of the axle of the gyroscope. Friction retards the precession, and the center of mass gradually falls until eventually the wheel hangs straight down.

HAZARDS

A spinning bicycle wheel is unwieldy and hard to control because of the unique properties of the gyroscope. You can stop the rotation by touching the wheel against something (shirt not recommended!). Be careful not to let the wheel graze your chest. Dizziness can onset quickly on the rotating stool or platform. It can induce sickness and cause you to fall after getting off the stool. Pause for a moment to regain your equilibrium before stepping off the stool. Be careful not to catch your fingers in the spokes of the wheel. If you use a volunteer on the rotating platform, locate the platform far from any objects that could cause injury if the person falls, and stand nearby to assist if the volunteer loses his or her balance.

REFERENCES

1. H. W. Dosso and R. H. Vidal, *Am. J. Phys.* **30**, 528 (1962).
2. J. R. Prescott, *Am. J. Phys.* **31**, 393 (1963).
3. C. T. Leondes, *Scientific American* **222**, 80 (Mar 1970).
4. J. S. Miller, *Physics Fun and Demonstrations*, Central Scientific Company: Chicago (1974).
5. G. D. Beadle, *Phys. Teach.* **27**, 488 (1989).
6. N. R. Greene, *Phys. Teach.* **35**, 431 (1997).
7. A. Bryant, *Phys. Teach.* **38**, 476 (2000).
8. R. H. Johns, *Phys. Teach.* **36**, 178 (1998).
9. R. H. Johns, *Phys. Tech.* **41**, 516 (2003).
10. H. F. Meiners, *Physics Demonstration Experiments*, Vol I, The Ronald Press Company: New York (1970).
11. D. E. H. Jones, *Physics Today* **23**, 34 (Apr 1970).
12. S. S. Wilson, *Scientific American* **228**, 81 (Mar 1973).
13. R. G. Hunt, *Phys. Teach.* **27**, 160 (1989).
14. J. D. Nightingale, *Phys. Teach.* **31**, 244 (1993).
15. L. A. Bloomfield, *How Things Work: The Physics of Everyday Life*, John Wiley & Sons: New York (2001).
16. H. A. Daw, *Am. J. Phys.* **56**, 657 (1988).

1.17
Stack of Cards

A stack of cards illustrates the static equilibrium of a rigid body by showing an impressive overhang.

MATERIALS

- stack of cards, blocks, bricks, books, or meter sticks

PROCEDURE

For a small audience, use a deck of ordinary playing cards. For a larger group, use a few dozen identical squares of cardboard, blocks of wood, bricks, or books to provide better visibility [1–3]. You can do a quantitative demonstration with a stack of meter sticks, which you could also suspend in the form of a mobile [4, 5]. Slide the uppermost card of the deck horizontally until it is just about to tip over (half of it must be supported by the card underneath). Then slide the top two cards in the same direction, and so forth, until the uppermost cards hang out a surprisingly large distance beyond the base of the deck. You can mark the proper position on the cards with a felt-tip pen. For more drama, let the stack hang out over the edge of the table [6] and have someone sit underneath. Then leave the audience to puzzle over why the deck does not topple.

Try pushing the upper-most card or block at right angles to the overhang so that it rotates though a small angle about its point of support. Then repeat the process with each card or block below it until you produce a freestanding spiral staircase. Point out that this illustrates the effect of a shear force, which changes the shape of an object but not its volume.

You can make an impressive, freestanding arch with a pair of card decks placed side-by-side and sheared in opposite directions. The Gateway Arch in St. Louis is constructed in this way. An object with a uniform weight per unit

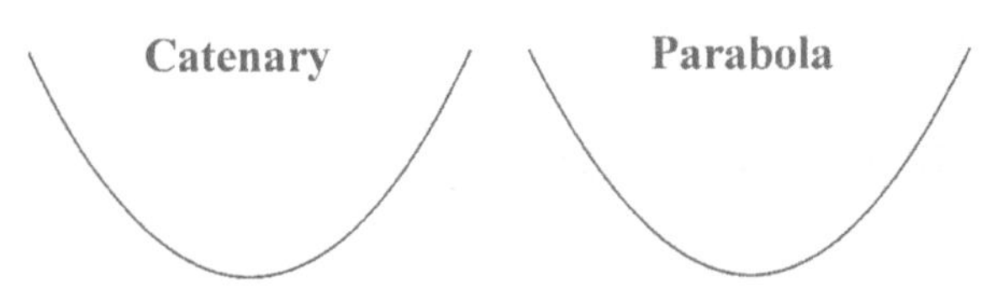

length will support itself if it has the shape of a catenary, $y = \cosh x$, which is the shape that would be produced by a uniform wet rope hung loosely between two poles, frozen, and then turned upside down. You could make such a rope on a sufficiently cold day. By contrast, the shape of a cable supporting a horizontally uniform weight, such as a suspension bridge, is a parabola, $y = x^2$.

Another way to demonstrate the principle of static equilibrium is to pick a volunteer from the front row, preferably someone who looks large and strong. Ask the person if he thinks you can prevent him from standing up using only one finger. When he looks skeptical, press your finger against his forehead so that he cannot lean forward. By preventing him from putting his center of gravity over his feet, he cannot stand up. He may try to swivel his hips forward, but a chair with arms on the side will prevent such movement although the armrest will allow him to push himself up with his arms. You can use your smallest finger for added effect. You can also show that if you stand with your back against a wall, you cannot bend over and touch your toes since the center of gravity has to be behind the tips of your toes. You can also do this demonstration with a volunteer.

DISCUSSION

The rule is that an object is in stable equilibrium if its center of gravity is above a point of support so that there is no gravitational torque on the object [7, 8]. With the stack of cards, the center of gravity of all the cards above a certain point must lie over a part of the card just beneath that point. The center of gravity of the top card is at its middle, and so it can overhang by one-half of its width. When so placed, the center of gravity of the top two cards is such that the second card can overhang by one-fourth of its width. The third card can overhang by one-sixth of its width, the fourth by one-eighth, and so on (see the diagram). Thus the total overhang of the top card in a deck of N cards is determined by the summation $1/2 + 1/4 + 1/6 + 1/8 + \cdots + 1/2N$, which is a harmonic series with a sum equal to $0.5(\gamma + \ln N)$ for large N. The quantity $\gamma = 0.577215662\ldots$ is the famous, but somewhat mysterious, Euler's constant (or Mascheroni's constant), which arises in Feynman diagrams among many other places [9]. Thus you can get an arbitrarily large overhang by using a sufficiently large number of cards. With 52 cards, the maximum stable overhang is 2.27 times the width of a card. In the presence of surface tension, which can be rather large if the cards have smooth surfaces, you can achieve an even greater overhang than suggested by the above formula. For the top card to overhang the bottom card completely, at least five cards are required.

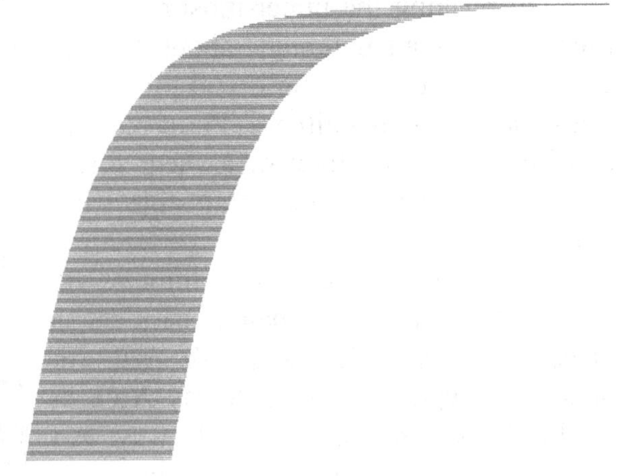

HAZARDS

There are no significant hazards with this demonstration unless you scale it up to such a large size that the weight of the pieces makes them dangerous if they topple.

REFERENCES

1. P. Johnson, *Am. J. Phys.* **23**, 240 (1955).
2. R. Sutton, *Am. J. Phys.* **23**, 547 (1955).
3. R. Boas, *Am. J. Phys.* **41**, 715 (1973).
4. R. Ehrlich, *Phys. Teach.* **23**, 489 (1985).
5. I. MacInnes, *Phys. Teach.* **27**, 42 (1989).
6. J. P. VanCleave, *Teaching the Fun of Physics*, Prentice-Hall: New York (1985).
7. L. Eisner, *Am. J. Phys.* **27**, 121 (1959).
8. G. K. Horton, B. Holton, and E. Freidkin, *Phys. Teach.* **35**, 214 (1997).
9. J. Havil, *Gamma: Exploring Euler's Constant*, Princeton University Press: Princeton, NJ (2003).

1.18
Coupled Pendulums

A rubber band near the top connects two rigid pendulums with the same resonant frequency, causing the energy to transfer back and forth between the two.

MATERIALS

- two identical rigid pendulums supported by a bar across the top
- rubber bands or weak springs

PROCEDURE

With the rubber band removed, show that the two pendulums are identical with the same period and that they each swing independently of the other. Attach the rubber band or spring between them, and bring both pendulums to rest. Then start one pendulum swinging in a plane perpendicular to the rubber band. After a while, the first pendulum will stop swinging, and the other will be swinging with large amplitude. Then the first will slowly begin swinging again while the second comes to rest, and so forth, until the energy damps away through friction. You can repeat the demonstration with rubber bands or springs of different stiffness connected at various distances from the pivot [1]. You can also use repelling magnets in place of the rubber bands or strings to couple the pendulums [2, 3]

DISCUSSION

The rubber band provides a weak coupling between the two pendulums allowing the energy to transfer slowly from one to the other. The stronger the coupling (stiffer rubber band, and/or farther from the pivot), the more rapidly the energy transfers. The phenomenon is the same as in the Wilberforce Pendulum (see section 1.19).

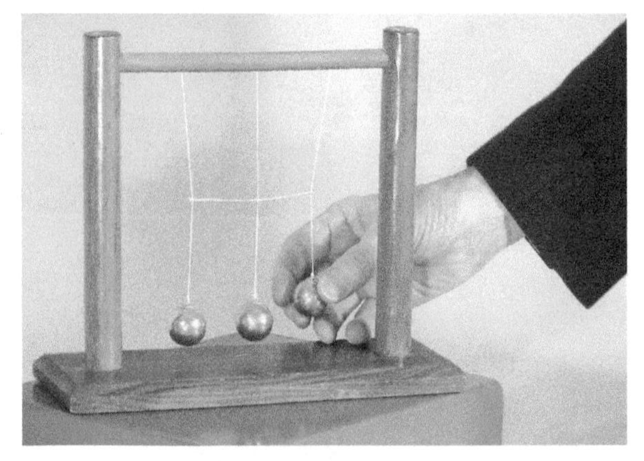

You can extend the idea to any number of pendulums, serving as an introduction to the motion of waves. With pendulums of various lengths, energy only transfers efficiently to those that are resonant with the one that it initially set into motion [4]. In electrical circuits, energy often transfers from one resonant circuit to another tuned to the same frequency by means of weak electrical coupling (see section 4.6). For radio and television transmission, the transmitter and receiver must be tuned to the same frequency and have narrow bandwidths.

HAZARDS

There are no significant hazards with this demonstration.

REFERENCES

1. J. S. Miller, *Physics Fun and Demonstrations*, Central Scientific Company: Chicago (1974).
2. C. A. Sawicki, *Phys. Teach.* **36**, 417 (1998).
3. C. A. Sawicki, *Phys. Teach.* **39**, 172 (2001).
4. J. Pizzo, *Phys. Teach.* **30**, 275 (1992).

1.19
Wilberforce Pendulum

A spring pendulum constructed such that the torsional and longitudinal frequencies are nearly identical slowly transfers its energy back and forth between the two modes of oscillation.

MATERIALS

- spring pendulum with specially constructed mass[1]

PROCEDURE

You can demonstrate the phenomenon of resonance with a simple but clever device called a Wilberforce pendulum [1–9], named after Lionel Robert Wilberforce (1861 – 1944), a demonstrator at the Cavendish Laboratory in Cambridge, England, around the turn of the century. It consists of a mass suspended from above by a spring. Such a pendulum has three modes of oscillation: (1) the ordinary swinging mode, (2) an oscillation along the axis of the spring, and (3) a torsional (twisting) mode. If the resonant frequencies of the second two modes are nearly identical and one mode is initially excited, the other mode will slowly acquire energy, and the energy will slowly transfer back and forth between the modes.

DISCUSSION

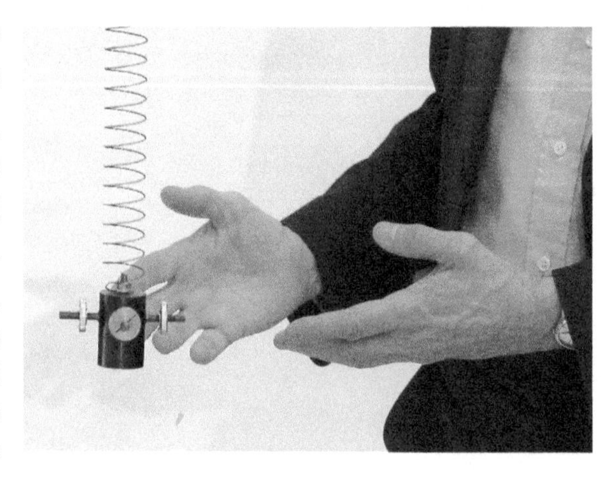

The angular frequency of the swinging mode is $\sqrt{g/L}$, where $g = 9.8$ m/s^2 and L is the length of the pendulum. The frequency of the spring oscillation is $\sqrt{k/m}$, where k is the spring constant and m is the mass supported by the spring. Finally, the torsional frequency is $\sqrt{K/I}$, where K is the torsional constant and I is the moment of inertia of the suspended mass. Usually bolts threaded into the mass in a symmetric arrangement control the

[1] Available from Educational Innovations and PASCO Scientific.

moment of inertia. Nuts threaded on the bolts can then be moved back and forth to change the moment of inertia without altering the mass. In such a way, you can make the two frequencies nearly equal.

If there were no coupling between the modes, the energy in each mode would remain constant, ignoring friction, and the modes could be excited in any combination with no subsequent interaction. In reality, the stretching of the spring produces a small torque that excites the torsional mode. The torsional mode, in turn, alternately stretches and compresses the spring, exciting the spring mode. The necessity of having the frequencies nearly equal is so that the coupling between the modes is small, and thus the energy must be transferred over a number of cycles. The effect is quite impressive if you adjust the frequencies carefully. The pendulum is usually designed so that the swinging mode has a sufficiently different frequency so that it does not couple to the other modes. This is an example of a harmonic oscillator driven at its resonant frequency by a small external force.

There are many other examples of coupled oscillators [10–25]. Usually they consist of only two matched frequencies. A pair of adjacent pendulums (coupled pendulums) attached together near the top with an elastic band or weak spring is another common demonstration, and you can use a succession of such pendulums to introduce the idea of wave propagation (see section 1.18).

HAZARDS

There are no significant hazards with this demonstration. Be careful not to damage the spring by stretching it too much.

REFERENCES

1. L. R. Wilberforce, *Philos. Mag.* **38**, 386 (1894).
2. L. R. Wilberforce in *Harmonic Vibrations and Vibration Figures*, edited by H. C. Newton, Newton and Company Scientific Instrument Makers: London (1909).
3. R. M. Sutton, *Demonstration Experiments in Physics*, McGraw-Hill: New York (1938).
4. J. Williams and R. Keil, *Phys. Teach.* **21**, 257 (1983).

5. R. J. Whitaker, *Phys. Teach.* **26**, 37 (1988).

6. V. Kopf, *Am. J. Phys.* **58**, 833 (1990).

7. R. E. Berg and T. S. Marshall, *Am. J. Phys.* **59**, 32 (1991).

8. J. Pizzo, *Phys. Teach.* **30**, 275 (1992).

9. F. G. Karioris, *Phys. Teach.* **31**, 314 (1993).

10. H. A. Robinson, Ed., *Lecture Demonstrations in Physics*, American Institute of Physics: New York (1963).

11. P. L. Tea and H. Falk, *Am. J. Phys.* **36**, 1164 (1968).

12. A. E. Siegman, *Am. J. Phys.* **37**, 843 (1969).

13. H. F. Meiners, *Physics Demonstration Experiments*, Vol I, The Ronald Press Company: New York (1970).

14. M. G. Olsson, *Am. J. Phys.* **44**, 1211 (1976).

15. T. E. Clayton, *Am. J. Phys.* **45**, 723 (1977).

16. J. G. Lipham and V. Pollak, *Am. J. Phys.* **46**, 110 (1978).

17. F. Pinto, *Phys. Teach.* **31**, 336 (1993).

18. W. R. Mellen, *Phys. Teach.* **32**, 122 (1994).

19. L. Falk, *Am. J. Phys.* **46**, 1120 (1978).

20. L. Falk, *Am. J. Phys.* **47**, 325 (1979).

21. M. G. Rusbridge, *Am. J. Phys.* **48**, 146 (1980).

22. H. M. Lai, *Am. J. Phys.* **52**, 219 (1984).

23. P. A. Bender, *Am. J. Phys.* **53**, 1114 (1985).

24. B. J. Weigman and H. F. Perry, *Am. J. Phys.* **61**, 1022 (1993).

25. N. W. Preyer, *Phys. Teach.* **34**, 52 (1996).

1.20 Chaotic Pendulums

Various pendulums exhibit chaotic motion when subjected to a nonlinear restoring force.

MATERIALS

- two or more disk-shaped magnets
- steel ball on the end of a string
- tennis ball (optional)
- overhead projector or video camera and monitor (optional)
- metronome (optional)

PROCEDURE

Suspend a steel ball or other ferromagnetic object from a string so that it comes within about two centimeters from a pair of disk-shaped magnets about five centimeters apart [1, 2]. You can place the steel ball inside a tennis ball to make the motion seem more mysterious. You can assemble the whole apparatus on top of an overhead projector to improve visibility or using a downward pointing video camera and monitor or make plots of the trajectory using a spark generator [3]. When released, the pendulum exhibits chaotic motion [4]. The motion eventually stops with the ball over one of the magnets, but the trajectory depends sensitively on the starting point. Alternately, hide a magnet in the tennis ball, and suspend it over four magnets, arranged in a square, two with north poles pointing upward, and two with north poles pointing downward. Contrast the irregular motion of the pendulum with the regular, periodic motion of a metronome. Point out that you cannot repeat the experiment because even the smallest change in the initial condition will eventually result in completely different behavior. This sensitive dependence on initial conditions is the defining characteristic of chaos.

You can best do this demonstration after you discuss and contrast the periodic motion of an ordinary pendulum. Point out that the difference arises because the restoring force produced by the magnet is highly nonlinear, unlike the gravitational force whose tangential component is approximately proportional to the angle the pendulum makes with the vertical. Each magnet has associated with it a basin of attraction that determines which initial conditions will approach that magnet. The basin boundary is a fractal (see section 6.13). Other chaotic pendulum demonstrations are also possible such as the periodically driven pendulum or the double pendulum [4–13]. At least three companies[1] market such devices for lecture and laboratory use [10].

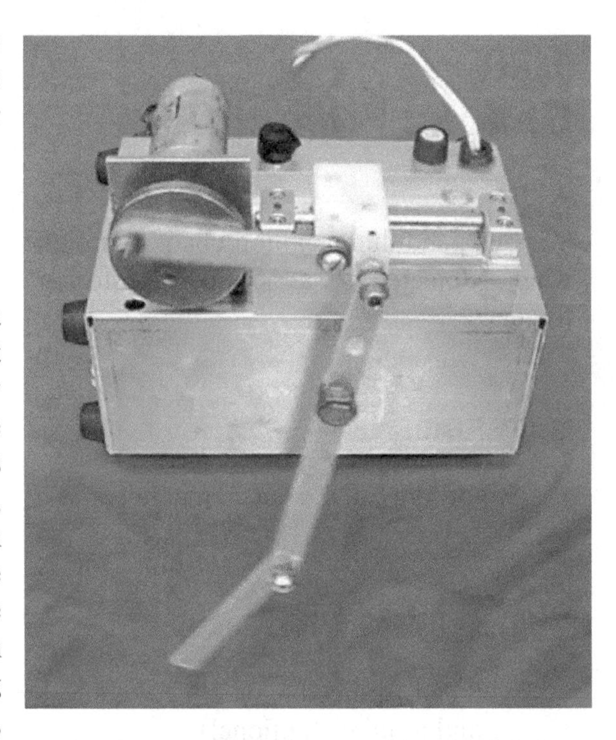

DISCUSSION

Chaotic motion arises whenever a deterministic system exhibits behavior that appears random and is sensitively dependent on the initial conditions. We now understand that even simple equations can have such solutions. The equations, however, must have a nonlinear term. In this case, the nonlinearity arises from the attraction of the magnetic dipole of the magnet to the induced magnetic dipole in the steel ball [14]. Such a force is strong but of a short range. With a single magnet, the motion would always decay to a unique final state, called a "stable equilibrium point." The second magnet allows two final states whose basins of attraction meet at a fractal boundary. The boundary exhibits structure no matter how greatly magnified. Thus two starting points near the basin boundary can have completely different final conditions. This sensitivity to initial conditions is called "the butterfly effect," because the atmosphere is presumably a chaotic, deterministic system, and thus a butterfly flapping its wings in Brazil, say, can set off tornadoes in Texas [15]. In fact, one of the earliest examples of chaos was discovered in a computer model of atmospheric convection [16], and the solution coincidently resembles a butterfly's wings. It is for this reason that long-term weather prediction is so difficult.

HAZARDS

There are no significant hazards with this demonstration.

REFERENCES

1. R. Ehrlich, *Turning the World Inside Out*, Princeton University Press: Princeton, NJ (1990).

2. D. Oliver, *Phys. Teach.* **37**, 174 (1999).

3. F. A. M. Cassaro, S. D. Saab, L. A. B. Bernardes, and J. B. da Silva, *Phys. Teach.* **42**, 47 (2004).

---

[1] Available from The Science Souorce, PASCO Scientific, and TEL-Atomic.

4. J. C. Sprott, *Chaos and Time-Series Analysis*, Oxford University Press: Oxford (2003).

5. G. L. Baker and J. P. Gollub, *Chaotic Dynamics: An Introduction* (2nd ed.), Cambridge University Press: Cambridge (1996).

6. N. Rott, *J. Appl. Math. Phys.* **21**, 570 (1970).

7. F. C. Moon, *Chaotic and Fractal Dynamics: An Introduction for Applied Scientists and Engineers*, Wiley-Interscience: New York (1992).

8. M. L. De Jong, *Phys. Teach.* **30**, 115 (1992).

9. A. Cromer and M. B. Silevitch, *Phys. Teach.* **30**, 382 (1992).

10. J. A. Blackburn and G. L. Baker, *Am. J. Phys.* **66**, 821 (1998).

11. J. P. Berdahl and K. Vander Lugt, *Am. J. Phys.* **69**, 1016 (2001).

12. R. DeSerio, *Am. J. Phys.* **71**, 250 (2003).

13. P. W. Laws, *Am. J. Phys.* **72**, 446 (2004).

14. H. O. Peitgen, H. Jürgens, and D. Saupe, *Chaos and Fractals: New Frontiers of Science*, Springer-Verlag: New York (1992).

15. E. N. Lorenz, *The Essence of Chaos*, University of Washington Press: Seattle (1993).

16. E. N. Lorenz, *J. Atmos. Sci.* **20**, 130 (1963).

1.21 Random Walk

By repeatedly flipping a coin to determine the direction to turn, one can execute a random walk.

MATERIALS

- coin
- computer with large monitor or projector (optional)
- a dozen or more Ping-Pong® balls in a large tray (optional)

PROCEDURE

Ask the audience if anyone has ever been on a random walk (sometimes called a "drunkard's walk"). Then offer to take them on one. Flip a coin. If it comes up heads, turn to the right and take a big step. If it comes up tails, turn to the left and take a big step. Repeat the process many times. Alternately, illustrate the method with several steps, and encourage people to go out into a large field and try it themselves. Another method that takes longer but works well when the streets in the neighborhood form a regular rectangular pattern like in Manhattan is to go a block between each flip of the coin.

Since this demonstration takes some time and gets boring after a few steps, it is ideal for a computer simulation. It is easy to program using the random number generator available in most computer languages. For example, a short BASIC program is the following:

```
10 SCREEN 1
20 CLS
30 X=160
40 Y=100
50 A=INT(4*RND)
60 IF A=0 THEN X=X+1
70 IF A=1 THEN X=X-1
80 IF A=2 THEN Y=Y+1
90 IF A=3 THEN Y=Y-1
100 PSET(X,Y)
110 GOTO 50
```

Tell the audience that this pattern is what you would see from an airplane flying high above someone who is executing a random walk. The pattern produced is a fractal (see section 6.13), and the trajectory will eventually return arbitrarily close to the starting point, but it might take a very long time [1]. The molecules in the air that we are breathing move in a random walk except in a three-dimensional rather than a two-dimensional space. They change direction when they collide with their neighbors, although the change in direction is not always a right angle, and the distance they go between collisions is not always the same. In three dimensions, a molecule never returns to its starting point.

Another way to demonstrate a random walk is to place a dozen or more Ping-Pong® balls in the bottom of a large tray. One of the balls should be a different color from the others. Rock the tray back and forth while viewing from above. The odd colored ball will execute a random walk among the other balls. In fact, all the balls are executing random walks, but it is easier to focus on one at a time.

DISCUSSION

Although the laws of classical physics are deterministic, when there are many particles interacting with one another such as in a gas at atmospheric density, the number of equations is very large, and it is a practical impossibility to solve them all. In such cases, statistical methods are more appropriate, and particles in such a collection will move in an effectively random manner. If the particles behave like hard spheres, as they tend to do if they are electrically neutral, they move in straight lines until a collision occurs (short-range interactions), whereupon the particles deflect off one another in arbitrary directions. Mechanical energy is usually conserved in the collision, although energy may transfer from one particle to the other. If the particles are electrically charged, the forces are long-range, and the deviation from a straight line is gradual and continuous, but the net result is the same. The same is true if the particles are instead massive astronomical objects such as stars or even whole galaxies governed by long-range gravitational forces.

A random walk is characterized by a distance *d*, called the "mean free path" (the average distance a particle goes before it experiences a collision), and a time τ, called the "collision time" (the average time between successive collisions). The ratio of the two is the speed of the particle, $v = d/\tau$. The average distance a particle moves from its initial position after a time t is $r = d\sqrt{t/\tau}$ (the mean free path times the square root of the number of collisions it has experienced) [2]. The reason for the square root becomes clear if one considers a particle that moves a distance *d* and then experiences a 90° collision and then moves another distance *d*. From the Pythagorean Theorem, the particle has thus moved a distance $d\sqrt{2}$ in a time $t = 2\tau$. Thus if you flip a coin a hundred times and take a step in a random direction each time, the average distance you go is ten steps. Do not expect this to work after a single trial, but the average of a large number of trials should approach the expected result.

2
Heat

Although heat at first appears to have nothing to do with motion, we now understand that heat is the random motion of molecules. The connection between heat and motion was provided by Benjamin Thompson[1] (1753 – 1814), an American who sympathized with the British during the Revolutionary War and eventually settled in Bavaria and became Count Rumford [1]. While watching the boring of artillery cannons in Munich in 1798, he concluded that motion can transform into heat and performed some experiments to demonstrate that fact.[2] Half a century later William Thomson[3] (1824 – 1907), later known as Lord Kelvin, used Rumford's data to measure the mechanical equivalence of heat. The experiments were done more carefully by James Prescott Joule[4] (1818 – 1889), the son of an English brewer. Joule's careful quantitative measurements became a model for modern science [2, 3]. This chapter on heat will also include such topics as fluid mechanics, phase transitions, cryogenics, thermodynamics, kinetic theory, and combustion.

REFERENCES

1. S. C. Brown, *Benjamin Thompson, Count Rumford*, MIT Press: Cambridge, MA (1979).

2. H. J. Steffens, *James Prescott Joule and the Concept of Energy*, Science History Publications: New York (1979).

3. T. B. Greenslade, *Phys. Teach.* **40**, 243 (2002).

[1] Benjamin Thompson also founded the Royal Institution in London.
[2] Now defined as 4.186 joules/calorie.
[3] William Thomson was a child prodigy who influenced the work of Joule and Maxwell, but he did not believe in atoms, and he opposed Darwin's theory of evolution and Rutherford's ideas of radioactivity.
[4] James Prescott Joule, more than anyone, is probably rightly credited for the concept of energy conservation and dispelling the caloric theory of heat, well before the existence of atoms was accepted. The unit of energy, the "joule," is usually pronounced "jool" even though Joule pronounced his own name as "jowl."

Table 2.1

Physical properties of selected gases

Gas	Formula	Molecular weight (g/mole)	Freezing point (°C)	Boiling point (°C)	Density (g/liter) @ 1 atm
Argon	Ar	39.95	−189.2	−185.9	1.633
Carbon dioxide	CO_2	44.01		−78.5[5]	1.799
Helium	He	4.00	−272.2[6]	−268.9	0.164
Hydrogen	H_2	2.02	−259.1	−252.9	0.082
Methane	CH_4	16.04	−182.6	−161.5	0.656
Neon	Ne	20.18	−248.7	−246.1	0.825
Nitrogen	N_2	28.01	−209.9	−195.8	1.145
Oxygen	O_2	32.00	−218.4	−183.0	1.308
Sulfur hexafluoride	SF_6	146.05		−63.7[7]	5.970
Xenon	Xe	131.30	−112.0	−108.1	5.367

[5] Sublimation point.
[6] At 26 atmospheres. Helium will not freeze below a pressure of about 20 atmospheres.
[7] Sublimation point.

General Safety Considerations

The demonstrations in this chapter present special hazards, since most involve very hot or extremely cold substances, volatile chemicals, fragile glassware, high-pressure gases, or evacuated containers. You should develop procedures to ensure that you can do the demonstrations without danger to the audience or to yourself. Handle hot and cold substances with tongs or with special, thermally insulated gloves. Wear eye protection not only to avoid personal injury but also to set a proper example for the audience. You should have an ABC-rated fire extinguisher, first aid kit, safety shower, and telephone available nearby.[1] You should know where these are located and how to use them. An assistant trained in first aid and CPR provides an extra measure of protection. You should especially caution children not to attempt certain of the demonstrations and not to do others except under the supervision of a parent or another adult who understands the physics principles. You should explain the reason for the precaution since such warnings sometimes have a contrary effect. You should have evacuation instructions clearly posted, planned, and tested.

In the event of a burn, promptly put the burned skin under cold water until the pain goes away. If blisters are present, cover them with sterile gauze. See a doctor if the burn is severe or if a large area is involved. With frostbite, warm the area slowly with lukewarm (about 106°F) water or with another part of the body, such as putting a finger under an armpit. Clean cuts and abrasions gently with soapy water, making sure dirt and other foreign material is flushed out, and then dry the area and cover it with a sterile bandage. If bleeding is severe, apply pressure to the cut and/or the pulse point above the cut. Elevate the cut above the level of the heart. If bleeding persists or if the cut is severe, see a doctor within a few hours to have the cut stitched.

Safety considerations specific to sound, electricity, magnetism, and lasers will be covered in subsequent chapters.

[1] Available from Lab Safety Supply, P. O. Box 1368, Janesville, WI 53547-1368, 608-754-2354.

2.1
Magdeburg Hemispheres

Two hemispheres when placed together and evacuated cannot be pulled apart because of the atmospheric pressure.

MATERIALS

- Magdeburg hemispheres[1]
- vacuum pump
- pressure gauge (optional)
- bathroom plungers (optional)

PROCEDURE

You can vividly demonstrate the large force that the atmosphere exerts using a pair of Magdeburg hemispheres, named after the city of Magdeburg in East Germany where Otto von Guericke (1602 – 1686), a physicist and politician, while mayor of the city in 1654, entertained the local people by having two teams of eight horses each try to pull apart two half-meter-diameter, copper hemispheres held together only by the pressure of the atmosphere [1, 2]. Otto von Guericke devised one of the first vacuum pumps in about 1641 and was largely responsible for dispelling the medieval view that "nature abhors a vacuum."

For a more modest demonstration, you can use steel hemispheres with a diameter of about 12 centimeters and a gasket of some sort to prevent leakage of air around the joint. If a good vacuum pump is not available, you can use larger hemispheres and a water aspirator or hand pump to produce a partial vacuum. It is even possible to lower the pressure to around 450 Torr (about 60% of atmospheric pressure) by sucking on the hose with your mouth.

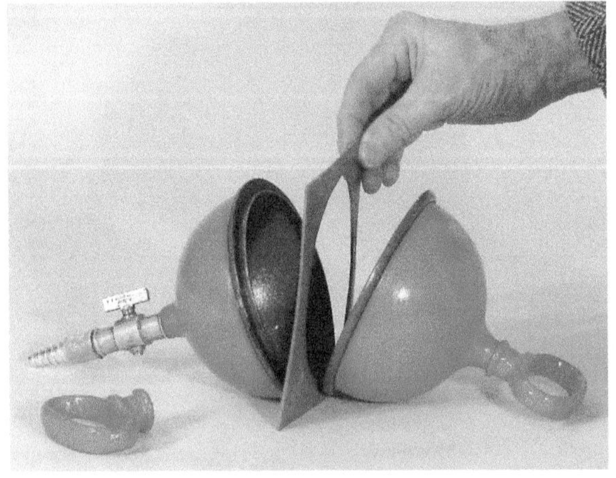

[1] Available from American 3B Scientific, Carolina Biological Supply Company, Fisher Science Education, Frey Scientific, PASCO Scientific, Sargent-Welch, Science First, and Ward's Science.

One of the hemispheres should have a lip around the edge to prevent the hemispheres from sliding across the gasket. Place the hemispheres together and evacuate them with a vacuum pump through a hose connected to a valve on the side of one of the hemispheres. Then shut the valve and remove the hose. Ask the audience what is inside the sphere. Common answers are "nothing," "air," or "vacuum," all of which are good answers. The hemispheres are usually equipped with handles so that two volunteers from the audience can try to pull them apart. It is safe to offer them a sum of money if they succeed (without opening the valve!).

In a variation of the demonstration, you can use two disks with the same projected area as the hemispheres to show that the volume of the evacuated space is not the relevant factor. You can show rubber suction cups of various types. A bathroom plunger is a familiar example. You can use a pair of such plungers as an inexpensive version of the demonstration.

DISCUSSION

The atmosphere exerts a force of $\pi r^2 p$ or 6.25π sq in $\times$ 14.7 pounds/sq in = 290 pounds for a 5-inch-diameter sphere completely evacuated. Even with only a partial vacuum, it is nearly impossible for two people of normal strength to separate them. You can hang weights from one of the hemispheres with the other attached to a secure hook in the ceiling to get a quantitative measure of the atmospheric pressure. With Otto Von Guericke's half-meter hemispheres, the force is over two tons! The same large pressure is pushing on everything, including people, but we do not normally notice it because we have an equal pressure inside our bodies pushing outward. Sometimes, you do notice it when you are in a rapidly moving elevator or airplane climbing or descending.

HAZARDS

Assuming the spheres are made of a material sufficiently strong not to be crushed by the atmosphere, the only danger is from one of the volunteers falling if the other pulls too hard or stops pulling. A suitable conclusion to the demonstration is to open the valve and let air in while the volunteers are still pulling, in which case you should instruct them not to pull too hard. Make sure the belt of the vacuum pump has a guard, or place the pump out of reach.

REFERENCES

1. J. Higbie, *Am. J. Phys.* **48**, 987 (1980).

2. H. Petroski, *Am. Sci.* **84**, 15 (1996).

2.2
Bernoulli Effect

A fluid such as air flowing over an object such as a balloon reduces the pressure above the object and levitates it.

MATERIALS

- hair dryer
- air-filled balloon or Ping-Pong® ball
- dinner roll, egg, beach ball, or empty plastic soda bottle (optional)
- spoon and water stream (optional)
- two sheets of paper (optional)
- glass funnel and Ping-Pong® ball (optional)

PROCEDURE

There are many ways to demonstrate the Bernoulli effect [1–8]. One of the simplest demonstrations involves a balloon filled with air or a Ping-Pong® ball and a hair dryer, which you use only to provide a jet of air, not for heating the air. Most hair dryers have a setting that allows the blower to run with the heat off. The balloon levitates a meter or so above the hair dryer and the Ping-Pong® ball a bit lower by the flow of air. You can try to blow the ball out of position with your mouth to illustrate its stability. The demonstration is most effective if you rigidly attach the hair dryer to the lecture bench and aim the jet of air at an angle of at least 30° from the vertical so that the audience does not assume that you are suspending the object using dynamic feedback. You can also do the demonstration in the dark with a spotlight or black light with fluorescent paint on the balloon and let the audience guess why it does not fall, perhaps with loud music to disguise the noise of the hair dryer [9].

You can try levitating other more amusing objects such as a dinner roll, an egg, a beach ball, or an empty plastic soda bottle. You can use a vertical stream of water (a fountain) in place of the air stream, but this variant is likely to get you wet and make a bit of a mess unless you have a good way to collect the water. With a fire hose, you can successfully levitate a basketball, football, or even a small watermelon, but this demonstration is best done outdoors.

An alternate demonstration involves suspending a spoon near a stream of water such that the convex side of the spoon attracts to the water. A curved strip of aluminum foil will also work. A similar effect occurs when the stream of water from a shower causes the shower curtain to bow inward, although air convection also plays an important role [10]. Air flowing between two sheets of paper suspended a few centimeters apart will cause them to pull together in contrast to most people's intuition. You can demonstrate this effect by blowing gently between the pieces of paper. Yet another method is to place a Ping-Pong® ball in a glass funnel and blow into the stem of the funnel. The ball is attracted to the funnel rather than expelled, and you should be able to turn the funnel upside down without having the ball fall out.

DISCUSSION

The Bernoulli effect was first formulated by Daniel Bernoulli (1700 – 1782) in 1738. Daniel Bernoulli was one of a large Swiss family of scientists and mathematicians in which there was unfortunate rivalry, jealousy, and bitterness. The Bernoulli effect is a consequence of Bernoulli's law

$$p + \rho v^2/2 + \rho g h = \text{constant}$$

where p is the pressure of the fluid, ρ is the density, v is the flow velocity, h is the height, and g is the acceleration due to gravity (9.8 m/s^2). The simplest interpretation of Bernoulli's law is that it is a statement of the conservation of energy for an infinitesimal volume of fluid. The first term is the negative of the work done on the

fluid by the surrounding fluid elements, the second term is the kinetic energy, and the third term is the gravitational potential energy, which is essentially constant in this demonstration. Thus the pressure is smallest when the flow speed is largest.

If the air were flowing equally above and below the balloon, there would be no net force on the balloon due to the Bernoulli effect. However, the balloon would then fall due to its weight until it reaches a point where most of the flow is over the top of the balloon, reducing the pressure there. The moving air exerts an additional force on the balloon in the direction in which it flows. The balloon sits at a point of stable equilibrium, where these three forces balance. The Bernoulli effect is one of the sources of lift when air flows over the top of the wings of an airplane, but the angle of attack is at least equally important, allowing airplanes to fly upside down [11–14]. In fact, the two explanations can be considered as two descriptions of the same phenomenon when certain simplifying assumptions are removed [15].

HAZARDS

There are no significant hazards with this demonstration.

REFERENCES

1. A. Sieradzan and W. Chafee, *Phys. Teach.* **27**, 306 (1989).
2. J. Pizzo, *Phys. Teach.* **27**, 308 (1989).
3. R. P. Bauman and R. Schwaneberg, *Phys. Teach.* **32**, 478 (1994).
4. B. Holmes, *Phys. Teach.* **34**, 362 (1996).
5. H. Cohen and D. Horvath, *Phys. Teach.* **41**, 9 (2003).
6. J. S. Miller, *Demonstrations in Physics*, Ure Smith Publishers Ltd.: Sidney (1969).
7. H. E. Meiners, *Physics Demonstration Experiments*, Ronald Press Company: New York (1970).
8. R. M. Sutton, *Demonstration Experiments in Physics*, McGraw-Hill: New York (1938).
9. M. Magalhães and F. Saba, *Phys. Teach.* **35**, 294 (1997).

10. A. A. Bartlett, *Phys. Teach.* **34**, 444 (1996).

11. D. Dommasch, S. Sherby, and T. Connolly, *Airplane Aerodynamics* (4th ed.), Pitman Publishing Company: New York (1967).

12. N. Smith, *Phys. Teach.* **10**, 451 (1972).

13. K. Weltner, *Am. J. Phys.* **55**, 50 (1987).

14. C. Waltham, *Phys. Teach.* **36**, 457 (1998).

15. C. N. Eastlake, *Phys. Teach.* **40**, 166 (2002).

2.3
Bell Jar

Objects placed in a bell jar connected to a vacuum pump expand when the jar is evacuated and contract when air is readmitted.

MATERIALS

- bell jar[1]
- vacuum pump
- balloon
- water, carbonated beverage, marshmallows, shaving cream, or doll head (optional)

PROCEDURE

Inflate a balloon with air to about half the diameter of the inside of the bell jar, and tie it off. Place it inside the bell jar. Ask the audience to predict what will happen when you remove the air from the bell jar. Some will say it will get smaller, some will say bigger, others will say it will pop, and still others will say it will float. Turn on a vacuum pump to evacuate the bell jar. The balloon expands. Then ask the audience to predict what will happen when you readmit air to the bell jar. Turn the pump off and readmit the air, causing the balloon to return to near its original size. You can ask the audience why that happened.

You can repeat the procedure with other objects such as a glass of water, a carbonated beverage, marshmallows, shaving cream, or a doll head [1–3]. When pumping on water, use a sulfuric acid trap to prevent the pump oil from becoming contaminated. You might also loosely cover the exhaust hole with a cup to prevent the hole from clogging or objects from entering the pump. Some of these demonstrations tend to be messy and require cleanup.

DISCUSSION

The size of a balloon depends on the pressure difference between the internal gas and the external gas as well as the elastic tension of the balloon material [1]. The balloon expands when you reduce the external pressure and contracts when you increase the

---

[1] Available from American 3B Scientific, American Scientific Products, Carolina Biological Supply Company, Fisher Science Education, Frey Scientific, Nasco, Sargent-Welch, and Ward's Science.

external pressure. A balloon with a diameter of twelve inches at sea level has a total external force in excess of three tons! Consequently, even a partial vacuum will significantly change its size. The balloon may not return to quite its original size because the balloon material may permanently stretch when it expands. It is for this reason that a balloon is easier to inflate the second time than the first. A balloon is also easier to inflate if you first stretch it.

A marshmallow resembles a large number of tiny balloons. Most people know that a marshmallow is mostly air. If you leave the marshmallow in the evacuated bell jar for a few minutes, the air will mostly leak out of the trapped pockets, and the marshmallow compresses to about half its original size when you readmit air quickly enough that the pockets do not refill. With a carbonated beverage, the carbon dioxide comes out of solution because of the reduced pressure of the external gas. The liquid will begin to boil if you leave the pump on too long (see section 2.7).

HAZARDS

The main hazard with this demonstration is implosion of the bell jar, which has many tons of atmospheric pressure on it while evacuated. Use only bell jars designed to be evacuated, and be careful not to drop the bell jar or strike it with any hard objects. Make sure the belt of the vacuum pump has a guard, or place the pump out of reach.

REFERENCES

1. J. Walker, *Scientific American* **261**, 136 (Dec 1989).
2. J. D. Wilson and John Kinard, *College Physics*, Allyn and Bacon: Boston (1990).
3. A. DePino, *Phys. Teach.* **39**, 56 (2001).

2.4 Collapsing Can

A small amount of water placed in an aluminum soft drink can is brought to a boil over a Bunsen burner, and then the can is inverted in a bath of cold water, causing it to collapse instantly from the rapid condensation of the steam.

MATERIALS

- aluminum soft drink can
- 15 ml of water
- Bunsen burner or hot plate
- large tongs
- tray of cold water
- safety glasses
- ice cubes (optional)
- tray of boiling water (optional)
- tray of liquid nitrogen (optional)
- 55-gallon drum (optional)
- 2-liter plastic soda bottle (optional)

PROCEDURE

Pour about 15 milliliters of water into an empty, 12-ounce, aluminum, soft drink can (one with "Crush" in its name is good) and heat it over a Bunsen burner or hot plate until a cloud of condensed water vapor escapes from the mouth of the can for about 20 seconds. Point out that the cloud is not "steam," which is an invisible gas, nor is it smoke, which is a collection of tiny solid particles. Since it takes several minutes for the steam to form and fill the can, you might want to do something else while waiting.

Using tongs, quickly lift the can from the burner, and invert it in a tray containing cold water to a depth of a few centimeters. The can will collapse instantly [1–5]. A few

ice cubes in the water makes the effect even more dramatic. You can heat several cans at once so that the demonstration can be repeated without additional delay. Point out to the audience that they can do this experiment at home using a stove in place of the Bunsen burner and hot pads in place of the tongs.

In a variant of the demonstration, invert the can over a tray of boiling water [6]. The can will not crush because the thermodynamic efficiency (the efficiency of converting heat into useful work of crushing the can) approaches zero when the hot and cold reservoir are at the same temperature. In a heat engine, heat has to flow from a hot body to a cold body to do work according to the second law of thermodynamics, which states that the total entropy[1] of the universe cannot decrease. If you invert the can over a tray of liquid nitrogen, the can will not crush, presumably because it fills with nitrogen gas from the boiling liquid and the gas does not conduct heat efficiently from the steam [5].

There are more dramatic versions of this demonstration using much larger containers such as a 55-gallon drum in which you put some water and bring it to a boil with the container open to the air and then cap off the container and either wait for it to cool or spray cold water on it. To make it collapse even more rapidly, cool it with ordinary ice, dry ice, or liquid nitrogen. For a simpler demonstration, pour a bit of hot tap water, or better yet, boiling water, into a 2-liter plastic soda bottle, shake it vigorously, empty the bottle, and quickly cap it, after which it will gently buckle inward as the vapor pressure of the water decreases [7].

DISCUSSION

When the water boils, it fills the can with steam, displacing most of the air that was originally in the can. When the can cools, the steam condenses back into liquid, reducing the pressure inside the can from which most of the air has been expelled. The surface area of the can is about 0.031 m^2 (or 48 in^2). In the extreme case, a pressure difference of one atmosphere (14.7 lb/in^2) would apply a total force of about 700 pounds, which is far in excess of what the can is able to sustain. You might expect the water to rise up into the can. However, since the hole in the can is small and the condensation occurs quickly, the inertia and viscosity of the water inhibits it from entering the can. Some water does spray up into the can, and this water causes the steam to condense rapidly.

---

[1] The entropy is a measure of the disorder in a system. It has been estimated that the total decrease in entropy resulting from all human activity since the dawn of humanity is offset by the increase in entropy resulting from striking a match [8].

HAZARDS

The can contains boiling water at 100°C and can cause burns to the skin if it is touched with unprotected hands. You should wear safety glasses. Be sure the outside of the can is clean and the can contains only water to prevent it from catching fire when you apply heat. If you scale the demonstration up to a larger size, the can collapses with considerable noise and force and should be done away from the audience, behind a safety screen, or viewed by video.

REFERENCES

1. G. Kauffman, *Journ. College Sci. Teach.* **14**, 364 (1985).

2. B. Z. Shakhashiri, *Chemical Demonstrations*, The University of Wisconsin Press: Madison, WI, Vol 2 (1985).

3. P. G. Hewitt, *Phys. Teach.* **26**, 491 (1988).

4. R. D. Edge, *Phys. Teach.* **29**, 144 (1991).

5. J. E. Stewart, *Phys. Teach.* **29**, 144 (1991).

6. B. W. Holmes, *Phys. Teach.* **35**, 281 (1997).

7. J. Holzman, *Phys. Teach.* **29**, 429 (1991).

8. R. Mills, *Space, Time and Quanta: An Introduction to Contemporary Physics*, W. H. Freeman: New York (1994).

2.5
Hero's Engine

A specially constructed, glass flask containing water and suspended from above by a chain spins rapidly when heated from below.

MATERIALS

- flask with steam tubes (Hero's engine)[1]
- Bunsen burner
- matches
- water
- supporting chain and stand
- safety glasses
- lawn sprinkler with compressed air (optional)

PROCEDURE

A glass flask with glass tubes connected to the neck and bent to one side, suspended by a bead chain, containing a small amount of water, placed above a Bunsen burner, spins rapidly as the water boils. The glass tubes act as jets releasing the steam in one direction. A bead chain provides freedom to rotate since it will not wind up. Alternately, you can use a chord that winds up until it eventually stops the rotation, and the flask starts rotating in the opposite direction, making a torsional pendulum. The amount of water used is not critical. Enough should be used so that it does not all boil away, but if you use too much, it will take a long time to begin boiling. You can adjust the speed of rotation by varying the amount of heat provided from below. You can construct a more rugged version using an aluminum cylinder, but then the boiling is not visible [1].

You can introduce this demonstration by pointing out that the flask contains water and asking what will happen when you light the burner underneath. As the water begins to boil, ask what is coming out the nozzles. Most people will say steam, but you can point out that steam is an invisible gas and that what they are seeing are tiny droplets of liquid

[1] Available from Carolina Biological Supply Company, Fisher Science Education, Frey Scientific, and Sargent-Welch.

water that form when the steam comes into contact with the cool air and condenses back into liquid. Ask what will happen as more steam comes out faster. If you time your question right, the flask will begin rotating on cue, usually after about a minute of heating.

The device is called "Hero's engine" after the Greek engineer Hero of Alexandria[2] (ca. AD 10 – AD 70), who in the first century AD wrote about it but apparently never made one. It is also called an "eolipile." Ask the audience whether such an engine would work well in an automobile. You can make an analogy to a rotating lawn sprinkler, which you can demonstrate indoors using compressed air in place of water. If you conceal the fact that you are using air, the audience will instinctively duck when the lawn sprinkler is turned on, assuming they are about to get wet (see section 1.13).

DISCUSSION

This demonstration illustrates phase transitions, the conservation of angular momentum, Newton's third law of motion (action and reaction), and the principle of a rocket engine. As the water vapor is expelled, it imparts an equal and opposite momentum to the flask, causing it to rotate in the opposite direction. It does not work by pushing against the surrounding air, and it would work perfectly well in a vacuum. It illustrates the transformation of heat energy into mechanical energy and shows the large amount of steam that a small amount of water can produce. Note the great simplicity of this engine as compared to other forms of steam engines that have many complicated moving parts. Perhaps it would be more widely used were it not for its enormous inefficiency in converting heat into motion because the exhaust velocity is relatively low and much heat is lost (exhausted to the air). By contrast, the gasoline engine has a maximum efficiency of about 30%, and the diesel engine has a maximum efficiency of about 40%.

HAZARDS

The flask should be of the type designed to be heated with a burner. Be careful not to boil the water too rapidly lest the flask begin to spin uncontrollably. Mount it some distance from its vertical support to prevent breakage if it spins too rapidly. If the glass breaks, it can be ejected with some force, and so you should wear safety glasses. Extinguish the flame before the water has completely boiled. Allow the flask to cool before touching it with unprotected hands.

REFERENCE

1. L. Hirsch, *Am. J. Phys.* **46**, 773 (1978).

[2] The city of Alexandria was founded in 334 BC by Alexander the Great, a Greek military commander, but it came under Roman rule in 80 BC and is now the second largest city in Egypt.

2.6 Model Geyser

A model geyser consisting of a column of water heated from below with a Bunsen burner erupts periodically and shoots water up to the ceiling.

MATERIALS

- model geyser
- Bunsen burner

PROCEDURE

A working model of a geyser can be rather easily constructed [1]. It requires only a long, vertical, water-filled tube with a constriction at the top, a supporting structure, a Bunsen burner to heat the tube from below, and a catch basin to collect the water and return it to the tube after an eruption. Depending on the length and volume of the tube and the amount of heat used, the geyser will erupt every few minutes. A water column of about two meters height works well since it makes the boiling point about 5°C higher at the bottom than at the top. If the tube is made of Pyrex® glass, you can observe the activity in the whole geyser tube.

You can do the demonstration most effectively by lighting the burner before the lecture and not taking any notice of it until it first erupts. Then ask the audience at what temperature water boils. Many people will give the correct answer (100°C or 212°F). Point out that this value is correct only at sea level. Mention that water boils at a lower temperature at high altitude. It takes longer to cook an egg in Denver than it does at sea level. Also mention that water boils at a higher temperature in a pressure cooker.

Ask who has been to Yellowstone Park. Mention that Yellowstone Park was the first national park in the United States, established in 1872. Explain that natural geysers there and elsewhere work because water runs deep down into crevices in the Earth where the rock is hot enough to boil the water [2]. Contrast the turbulent nature of the boiling to the regular periodic nature of the eruptions. Ask if anyone knows the name of a famous geyser in Yellowstone and why they call it that. Old Faithful is not the largest or most regular geyser in the park, but it has become popular because it erupts more frequently than any of the other big geysers. It currently erupts on average every 74 minutes, but the intervals vary from about 45 to 110 minutes and are predictable (based on the duration of the previous eruption) to within about ±10 minutes with 90% confidence. It ejects water about 100 to 180 feet into the air, and the eruptions typically last for 1.5 to 5 minutes.

For added amusement, plant someone in the audience near the geyser to raise an umbrella when the geyser erupts. Explain the difference between a geyser and a volcano, since some people confuse the two.

DISCUSSION

The geyser operates by heating the water at the bottom of the tube to a superheated state (greater than the boiling point of water at atmospheric pressure). The water does not immediately boil, however, since the pressure of the water column above elevates the boiling point. When it finally does begin to boil, the pressure of the water column is relieved, and the boiling vigorously ensues, causing the eruption. The ejected water will cool upon contact with the air and with the catch basin and will run back down the tube, where the heating process begins again [3].

In addition to the eruption, it should be possible to observe tremors of various types as well as a water-hammer effect that occurs just after an eruption when the pressure of the steam momentarily prohibits the water from running down the tube. As the steam condenses, a partial vacuum forms in the geyser tube and audibly "sucks" the pool of water from the catch basin into the tube with great force. The resulting noise sounds like a belch and invariably elicits an amused reaction from the audience. You might say "excuse you" to get a laugh. You can produce a similar effect by pouring water into a funnel inserted through a rubber stopper into an empty flask. The same effect is the source of the clanking noise that occurs in plumbing when a portion of the water stream abruptly changes speed, producing a partial vacuum in the pipe.

HAZARDS

The geyser is relatively safe if you exercise normal caution in not touching the Bunsen burner or the tube containing the hot water. The water can cause burns if anyone is too close when it erupts, but by the time the water has sprayed into the air and fallen back down, it is no more dangerous than a hot shower. Some water may miss the catch basin and splatter onto the audience or damage other nearby equipment. The hot water and steam can damage the paint on the ceiling above the geyser. A glass geyser is fragile and easily broken, possibly causing cuts and burns. Be careful not to step on the hose to prevent the burner from going out and filling the room with natural gas. Carbon dioxide from the rapid warming of dry ice can also extinguish the flame since it falls to the floor. This is not a good demonstration to leave unattended since the eruption is unexpected and sudden and could startle someone.

REFERENCES

1. L. W. Anderson, J. W. Anderegg, and J. E. Lawler, *American Journal of Science* **278**, 725 (1978).

2. T. S. Bryan, *The Geysers of Yellowstone* (3rd ed.), University of Colorado Press: Boulder, CO (1995).

3. T. S. Bryan, *Geysers: What They Are and How They Work*, Roberts Rinehart: Lanham, MD (1990).

2.7
Freezing by Evaporation

Water at room temperature in a flask boils vigorously and then turns into ice when the pressure in the flask is reduced.

MATERIALS

- thick-walled, round-bottom boiling flask
- rubber vacuum hose with valve
- sulfuric acid trap
- vacuum pump
- distilled water
- pressure gauge (optional)
- video camera and projector (optional)
- liquid nitrogen (optional)

PROCEDURE

With the use of a mechanical vacuum pump, you can make water in a flask boil so vigorously at room temperature that the loss of heat due to evaporation causes the water to freeze [1]. You can purchase or construct the apparatus for such a cryophorus demonstration. If constructed, you should use a thick, round-bottom boiling flask and a trap that you can fill with concentrated sulfuric acid to isolate the flask from the vacuum pump. The acid absorbs the water vapor, thereby increasing the rate of boiling as well as preventing contamination of the pump oil by water. Distilled water works best. You can bring the water to a boil more quickly if you preheat it. A large pressure gauge visible to the audience is a useful addition. You can do a variant of the demonstration using a small drop of water sitting on a thin plastic membrane covering the top of a plastic foam cup inside a bell jar connected to the vacuum pump [2].

You can introduce this demonstration by asking the audience at what temperature water boils. Many people will answer with 212°F or 100°C. You should point out that this answer is correct only at atmospheric pressure (760 Torr), and that it is well known by the residents of high-altitude cities that it takes longer to cook a boiled egg there because water boils at a lower temperature at higher altitude. It is not accurate to say that it takes longer to *boil* an egg there because the water actually comes to a boil more quickly. You can point out that a boiling liquid tends to maintain a constant temperature, and that is why many foods are cooked in boiling water. In a pressure cooker, the increased pressure raises the boiling point.

Then turn on the vacuum pump and let the audience watch while the water boils furiously. Some of what appears to be boiling water is actually air that was dissolved in the water. You can touch the flask and point out that it is not hot at all, and, in fact, if anything, it feels a bit cool. While holding your hand on the flask, ask the audience at what temperature water will freeze. If you time your comments properly, the water will freeze on cue into a slushy form of ice. In a large hall, the audience can best see the ice in silhouette on a screen illuminated by an arc lamp or by a slide projector or with a video camera and projector. You can let air back into the flask, remove it from the vacuum system, and pass it around for everyone to see and feel.

You can do the same demonstration with liquid nitrogen [3]. Put about 100 ml of liquid nitrogen into the flask and wait for it to stop boiling. Then turn on the vacuum pump. The liquid will boil rapidly and then freeze (when the pressure drops to the triple point of 0.124 atmospheres) just as the water did. Most people, even scientists, have not seen solid nitrogen, but the South Pole of Triton, a satellite of Neptune, is covered with a vast plain of solid nitrogen at a temperature of 38K [4].

DISCUSSION

The barometric pressure drops by about 3% for each 1,000 feet above sea level. A city such as Denver, Colorado, is at an altitude of about 5,000 feet, and thus the average barometric pressure is around 650 Torr, and the boiling point of water is about 95°C. The lapse rate of the boiling point is about 3.33°C / km or 1.83°F / 1,000 feet [5]. On top of Mt. Everest (29,035 feet[1] or 8,850 meters) it is nearly impossible to make a pot of hot tea.

The heat of vaporization of water is 540 calories per gram, and the heat of fusion of water is 80 calories per gram. The density of water is 1 gram per milliliter. When the water evaporates, a large amount of heat leaves the water, causing the cooling. The evaporation of each milliliter of water is capable of freezing almost seven milliliters once the temperature of the water has decreased to 0°C. As the temperature drops, the vapor pressure of the water drops, and the boiling decreases greatly. This heat loss is why we feel cool when we get out of the bath or swimming pool. Animals perspire to keep cool by the same process, and the mechanism is very efficient. People have survived temperatures as high as 125°C for a period of time adequate to cook a steak [6]. If our perspiration evaporated too rapidly, we would freeze to death!

A simple way to demonstrate the large value of the heat of vaporization is to compare the time it takes to bring a given volume of water at room temperature (~20°C) to a boil (100°C) at atmospheric pressure with the time it takes to boil that water away. The time required to boil it away should be greater by approximately a factor of 540/80 = 6.75 if the heating rate is constant [7].

---

[1] The elevation of Mt. Everest was revised from its historical (1954) value of 29,028 feet on November 11, 1999, as a result of an expedition using the Global Positioning System, but the new value is not universally accepted.

HAZARDS

The major hazard is implosion of the flask if it is not designed to withstand atmospheric pressure or if it is dropped or struck with a heavy object. Sulfuric acid can also cause chemical burns and should be handled with great caution. Make sure the belt of the vacuum pump has a guard, or place the pump out of reach.

REFERENCES

1. H. A. Robinson, Ed., *Lecture Demonstrations in Physics*, American Institute of Physics: New York (1963).
2. M. Capitolo, *Phys. Teach.* **36**, 349 (1998).
3. T. C. Lamborn and R. Mentzer, *Phys. Teach.* **32**, 508 (1994).
4. J. I. Lunine, *Science* **261**, 697 (1993).
5. C. R. Bohren, *What Light Through Yonder Window Breaks?,* John Wiley & Sons: New York (1991).
6. P. Davidovits, *Physics in Biology and Medicine* (2nd ed.), Harcourt/Academic Press: San Diego (2001).
7. E. Linz, *Phys. Teach.* **33**, 294 (1995).

2.8
Boiling with Ice

Holding an ice cube or crushed ice against a sealed flask containing a mixture of hot water and steam causes the water to boil.

MATERIALS

- 2-liter, thick-walled, round-bottomed boiling flask
- rubber stopper
- Bunsen burner or hot plate
- ring stand and clamp or tripod stand
- ice cube or crushed ice
- two-hole rubber stopper (optional)
- thermometer (optional)
- glass tube (optional)
- rubber hose with pinch clamp (optional)
- tray to catch boiling water and melting ice (optional)

PROCEDURE

You can make water boil at a temperature well below 100°C by reducing the pressure of the atmosphere above the water [1–6]. Fill a 2-liter, thick-walled, round-bottomed boiling flask about a third to half way with water. Bring the water to a boil using a Bunsen burner or hot plate, and let it boil for about a minute to fill the flask with steam. Then turn off the burner and allow the water to cool for about a minute. When all the boiling has stopped, place a rubber stopper firmly on the flask. Hold an ice cube against the side of the flask closest to the steam, or pour crushed ice or ice water over it, causing the water to boil for about a minute.

Alternately, you can use a two-hole rubber stopper to seal the flask. Place a thermometer in one hole of the stopper. In the other hole, insert a glass tube, and connect it to a rubber hose that you can seal with a pinch clamp. Bring the water to a boil with a Bunsen burner or hot plate. Then allow the water to cool for one minute. Use the pinch clamp to seal the flask, invert the flask, and hold an ice cube against its bottom. The water in the flask will boil for about a minute. Be sure the stopper is secure, and hold the flask over a tray so that the hot water will not burn you if the stopper comes out.

DISCUSSION

The temperature at which a liquid boils depends on the pressure of the atmosphere above it. The water will boil when its vapor pressure (see table 2.2) exceeds the pressure of the surrounding gas. The vapor pressure increases approximately exponentially with temperature. In this demonstration, the pressure decreases below normal atmospheric pressure (760 Torr) by cooling the steam in the flask, causing the steam to condense. As the water boils, the pressure in the flask increases to the point where it is greater than the vapor pressure of the liquid, and the water stops boiling. The flask should be no more than half-full of water to provide space for the vapor produced by boiling. If you provide too little space, the vapor pressure of the liquid water may be reached before boiling is apparent. The goal is to cool the steam while letting the water remain as hot as possible. Eventually the water will cool to the point where its vapor pressure is below the pressure in the flask, and the boiling then stops.

HAZARDS

In addition to the hazard of burns from the Bunsen burner or hot plate and the hot flask, the flask may implode if you lower the pressure sufficiently. You can minimize this risk by using a thick-walled boiling flask rather than a standard round-bottomed one. It is safest to do the demonstration without inverting the flask. If you do invert it, be sure the heat is off and the boiling has completely stopped before inserting the stopper, and press the stopper in securely. You will need thermally insulated gloves to invert the flask, and you should prepare for the possibility of the stopper coming out and spilling the boiling water by placing a try underneath it and holding the flask by its neck so your hand will not get wet if the stopper comes out. Wear safety glasses with this demonstration.

REFERENCES

1. P. Joseph, F. Brandwein, E. Morholt, H. Pollack, and J. F. Castka, *A Sourcebook for the Physical Sciences*, Harcourt, Brace and World: New York (1961).
2. H. A. Robinson, Ed., *Lecture Demonstrations in Physics*, American Institute of Physics: New York (1963).
3. D. Baisley, *The Science Teacher* **47**, 45 (May 1980).
4. B. Z. Shakhashiri, *Chemical Demonstrations*, The University of Wisconsin Press: Madison, WI, Vol 2 (1985).
5. J. P. VanCleave, *Teaching the Fun of Physics*, Prentice-Hall: New York (1985).
6. V. Thomsen, *Phys. Teach.* **35**, 98 (1997).

Table 2.2

Temperature dependence of the vapor pressure of water

Temperature (°C)	Pressure (Torr)
0	4.579
5	6.543
10	9.209
15	12.788
20	17.535
25	23.756
30	31.824
35	42.175
40	55.324
45	71.88
50	92.51
55	118.04
60	149.38
65	187.54
70	233.7
75	289.1
80	355.1
85	433.6
90	525.76
95	633.90
100	760.000

2.9 Liquid Nitrogen

Objects placed in liquid nitrogen change their physical properties because of the reduced temperature.

MATERIALS

- Dewar flask of liquid nitrogen[1]
- plastic hose or rubber ball
- banana
- flower
- rubber ball or Ping-Pong® ball
- tongs
- gloves
- hammer
- glass dish
- balloon
- safety glasses
- egg and Teflon® pan (optional)
- small piece of cookie or a marshmallow (optional)

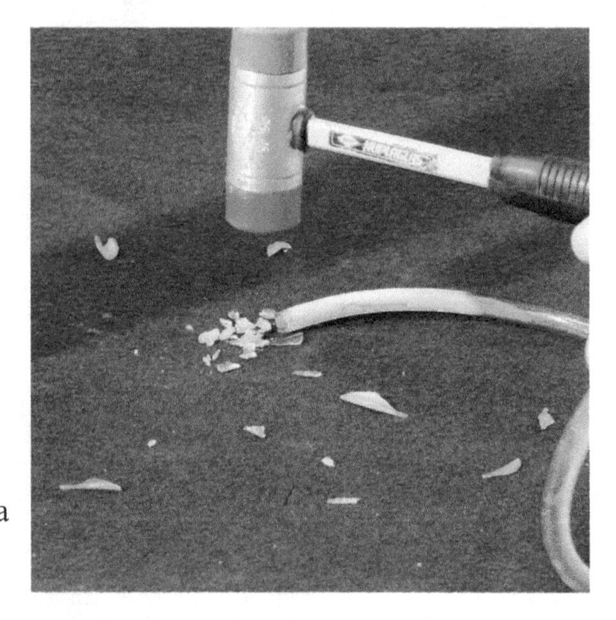

PROCEDURE

Pour a bit of liquid nitrogen onto the table, and ask the audience to note that the table did not get wet. The substance that looks like water is the major constituent (80% by volume) of the air that we breathe, but at a reduced temperature of −196°C (−321°F, 77K). It is probably the coldest substance most people have ever seen. Explain that it

[1] Liquid nitrogen is available from hospital supply houses. An ordinary stainless steel (not glass) thermos bottle serves as an adequate Dewar.

boils without evident heating because even ordinary materials like the table are much warmer than the temperature needed to boil nitrogen. Explain that cold is simply the absence of heat and that heat is not a substance but a form of energy. Mention that the vapor given off by the boiling is harmless since it just adds slightly to the nitrogen that is already in the air.

Explain that most materials change their physical properties when they get cold. Most people have noticed how hard the seats of their cars are on a cold morning. Dip a plastic hose into the Dewar flask of liquid nitrogen. It will make a fountain with liquid nitrogen spewing out the hose because of the boiling. While the nitrogen is boiling, a small cloud appears above the Dewar flask. Ask the audience what it is. Many will say "steam" or "smoke," but of course steam is invisible, and smoke is the product of combustion. The cloud consists of small particles of liquid water, about a thousandth of an inch in diameter, condensed from the air above the Dewar flask by the cooling effect of the boiling nitrogen. When the boiling stops, remove the hose and shatter it with a hammer.

Leave a banana in the liquid nitrogen for a few minutes and then use it to drive a large nail into a block of Styrofoam®. The banana has a tendency to shatter and leaves a mess when it thaws. In an extension of the demonstration, you can cool pieces of rubber, shaped like nails, and drive them into the Styrofoam® with the banana. With practice, it is possible to drive the nail into soft wood without shattering the banana. When you are done, shatter the banana by striking it on a hard surface if it has not already shattered. Clean up the pieces before they warm up.

Alternately, you can take a rubber ball, such as a squash ball, or a Ping-Pong® ball, and bounce it a few times on the floor and then drop it into the liquid nitrogen. When the boiling stops, remove the ball with tongs, and then, while wearing thermally insulating gloves, throw it to the floor, causing it to shatter. You could also dip a flower into the liquid nitrogen and then crumble it with your hand once it has cooled. Shattering a frozen orange is not recommended, since it leaves a sticky mess.

Inflate a balloon with air to about six inches in diameter, tie it off, and place it in a glass dish. Pour liquid nitrogen over the balloon, causing it to shrivel to negligible size. Then remove it from the dish with tongs or gloves and hold it while it slowly inflates again. If the balloon is clear and one looks carefully, it is possible to see some liquid air inside the balloon. Ask the audience to explain what is happening. Do the same thing with a helium-filled balloon. The balloon will float up to the ceiling when it warms up and expands.

Crack an egg into a Teflon® frying pan and lower the pan into the liquid nitrogen, fooling people into thinking you are boiling the egg. The cloud that rises from the pan resembles steam. The egg turns into slime when you allow it to warm up afterward.

Place a small piece of cookie ("biscuit" in the UK) or a marshmallow in a dish, and pour liquid nitrogen over it. When the boiling stops, remove the cookie or marshmallow with tongs, toss it into your mouth, and chew rapidly while exhaling. A cloud of condensed water vapor will come out your nose. Practice this demonstration with small pieces of cookie until you develop the confidence to do it safely.

DISCUSSION

The temperature of liquid nitrogen is sufficiently low that most materials have significantly different and sometimes surprising properties. Few elastic materials retain their elasticity at this temperature. Electrical resistance is much lower, and some materials in fact become superconductors and lose all electrical resistance at sufficiently low temperatures (see section 5.6).

When you cool the air in a balloon, its pressure decreases, and the pressure of the external atmosphere crushes it. Some of the nitrogen and most of the oxygen in the balloon may liquefy. Oxygen gas condenses into a liquid at a temperature of −183°C versus −196°C for nitrogen.

Liquid nitrogen is produced commercially by compressing air in stages to about 100 atmospheres while simultaneously cooling it to keep it near room temperature. Then the room-temperature gas expands through a nozzle, causing it to cool and liquefy. The liquid oxygen boils off, and the remaining liquid nitrogen is stored in a Dewar flask, which is a silvered, two-wall container with a vacuum between the walls, similar to a thermos bottle. Ordinary glass thermos bottles may shatter from the thermal stress if used with liquid nitrogen.

HAZARDS

Liquid nitrogen is potentially dangerous since it can cause instant frostbite. Avoid letting it come into contact with bare skin. Use tongs to lower objects into the liquid nitrogen and remove them. Be careful that the liquid does not splash into your eyes. You should wear gloves and safety glasses. When you shatter objects with a hammer or by throwing them against the floor or wall, the resulting projectiles can injure someone unless you take proper precautions. Ping-Pong® balls are much safer to shatter than squash balls, but less dramatic [1].

REFERENCE

1. K. Malone and T. Datta, *Phys. Teach.* **32**, 351 (1994).

2.10 Leidenfrost Effect

Liquid nitrogen poured over the hand causes no harm because of the heat capacity of the hand and the insulating layer of nitrogen vapor.

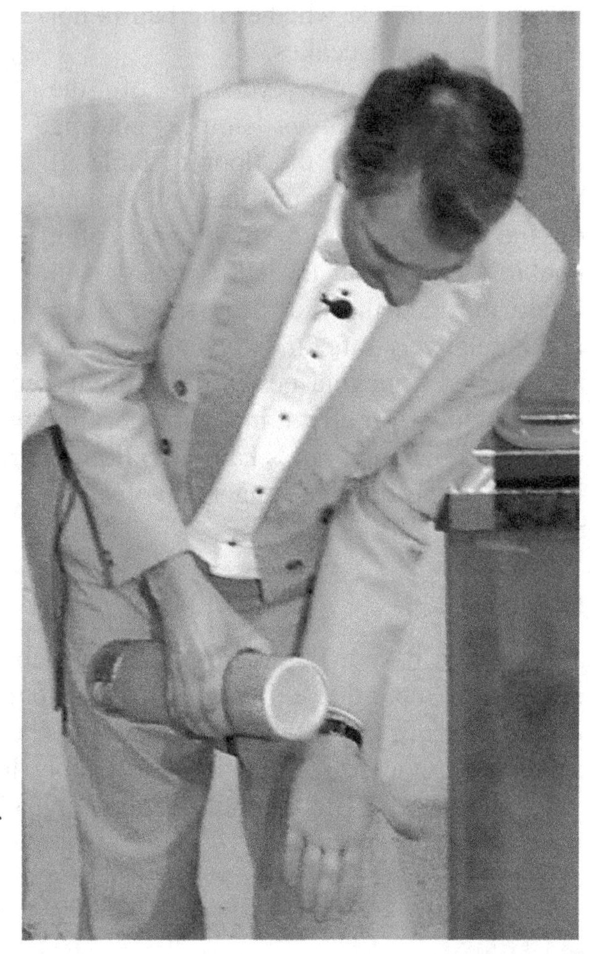

MATERIALS

- Dewar flask of liquid nitrogen
- hot plate and beaker of water
- safety glasses
- eyedropper (optional)
- skillet (optional)
- small piece of cookie (optional)

PROCEDURE

After demonstrating the effect of liquid nitrogen on various substances, ask the audience what would happen if you put your hand in the liquid nitrogen. Most people will realize that doing so would be dangerous, and assure them that they are correct. Then lift the Dewar flask and pour the nitrogen over your hand so that it runs off onto the floor. Ask why it is safe to do this. Tell them it is because of the "Leidenfrost effect" [1], named after Johann Gottlieb Leidenfrost (1715 – 1794), and that they are all familiar with it. Point out that knowing the name of something does not mean you understand it. Ask what happens when they sprinkle drops of water onto a hot skillet on the stove at home. Most people will recognize that the drops of water dance around without boiling at first. Explain that the water does boil but that a thin layer of steam forms where the water touches the skillet and the thermal conductivity of the steam is sufficiently low that not much heat transfers to the water.

With a hot plate and a beaker of water, you can demonstrate the effect to anyone who has not seen it at home. Dip your fingers in the water and let the drops fall onto the hot plate after it has thoroughly warmed. Alternately, use an eyedropper to release the drops. The drops will roll around on the surface of the hot plate until they fall off the edge. A skillet will prevent the drops from falling off, but unless it has low sides or the audience views it from above, it will be hard to see the drops. Turn the hot plate off when you are done. If the skillet is sufficiently hot, the drops will remain for a minute or longer. Point out that cooks sometimes use this method to test whether the pan is hot enough to cook pancakes.

An even safer version of the demonstration involves simply pouring liquid nitrogen onto the floor or table. It will form beads of liquid that roll around on the floor. Eating a small piece of a cookie ("biscuit" in the UK) cooled with liquid nitrogen illustrates the same effect and is relatively safe, and you can make a cloud of condensed water vapor from the cold air expelled from your nose after eating it (see section 2.9). Other popular demonstrations of the Leidenfrost effect, including gargling with liquid nitrogen, walking on a bed of hot coals, and plunging one's hand into liquid nitrogen or molten lead, are not recommended because of the extreme dangers involved.

DISCUSSION

Liquids have a Leidenfrost point, which is the temperature at which the boiling is sufficiently rapid that a drop of the liquid remains suspended on its vapor, in the manner of a hovercraft, rather than spreading out. The vapor layer has a typical thickness of 0.06 mm. The Leidenfrost point is considerably higher than the boiling point and is about 220°C for water. The thermal conductivity of gases is several orders of magnitude less than for liquids and solids.

HAZARDS

Be sure your hands are clean, and avoid cupping your hand when pouring the liquid over it. It is safer to use the palm rather than the back of the hand. Protect the hot plate so that no one can inadvertently come into contact with it while it is hot, and turn it off when you are done with the demonstration. Wear safety glasses when working with liquid nitrogen.

REFERENCE

1. J. G. Leidenfrost, *A Tract about Some Qualities of Common Water*, University of Duisburg: Duisburg, Germany (1756).

2.11 Liquid Nitrogen Cannon

The rapid evaporation of liquid nitrogen inside a steel cylinder exerts enough pressure to blow a cork stopper off the cylinder.

MATERIALS

- Dewar flask of liquid nitrogen
- steel cylinder with a smaller, internal, stainless steel cup with a wire handle
- cork stopper
- mallet
- safety glasses

PROCEDURE

A spectacular demonstration involving the vaporization of liquid nitrogen can be done by lowering a small stainless steel cup (~30 milliliters) suspended by a wire and filled with liquid nitrogen into a larger steel cylinder (40 cm long × 5 cm inside diameter), sealed on one end. Pound a cork stopper into the other end with a mallet. Then pick up and shake the whole device to spill the nitrogen from the cup. When the liquid nitrogen strikes the wall of the cylinder, it rapidly evaporates, causing a large and sudden pressure increase that blows the cork off with great force and noise. A heavy base on the cylinder helps absorb the shock of the recoil. Alternately, let the cylinder rest on the table in a vertical position until the nitrogen boils and expels the cork upward. You can have a volunteer or assistant try to catch the cork, perhaps using a fish net. This is a good opportunity to caution the audience that many of the things they will see are potentially dangerous and should be repeated only with the help of a parent, teacher, or other adult who understands the scientific principles involved.

You can procure liquid nitrogen from hospital supply sources. If a Dewar flask is not available, you can use a stainless steel (not glass) thermos bottle to store the nitrogen, but you should not seal it tightly lest the thermos bottle explode as the nitrogen evaporates. If liquid nitrogen is not available, you can instead use a piece of dry ice [1]. You can attach streamers to the top of the cork for more drama.

DISCUSSION

When a liquid evaporates, the resulting gas occupies a much larger volume than the liquid from which it came if the pressure is the same in both cases. The volume expansion is typically about a factor of 1,000 for most liquids (compare the density of the gas with the density of the liquid from which it came). If the gas cannot expand, the pressure will increase by the same factor. The increased pressure raises the boiling point and inhibits the boiling, as in a pressure cooker, but not enough to prevent a large pressure buildup, as demonstrated here. In the liquid nitrogen cannon, only a small fraction of the liquid nitrogen needs to boil for the cork to blow off with considerable force.

HAZARDS

Liquid nitrogen boils at a temperature of −196°C (−321°F, 77K) and can cause severe frostbite. Take care to avoid contact with the liquid and with anything that has been cooled by it. Aside from making sure the cylinder is quite strong, the only other precaution is to aim the cannon above the heads of the audience. The cork can travel several hundred feet, but it is relatively harmless after it bounces off a wall or the ceiling. If you propel it straight upwards, be sure there are no fragile lights on the ceiling. Do not use a rubber stopper because it is harder and heavier than the cork and thus more dangerous if it hits someone. Wear safety glasses when working with liquid nitrogen.

REFERENCE

1. J. S. Miller, *Physics Fun and Demonstrations*, Central Scientific Company: Chicago (1974).

2.12
Liquid Nitrogen Cloud

Liquid nitrogen induced to vaporize rapidly by expelling it from a large Dewar flask under pressure cools the air and causes the formation of a large, dense cloud.

MATERIALS

- 25-liter Dewar flask filled with liquid nitrogen
- 2-hole rubber stopper to fit the Dewar flask
- specially constructed copper pipe with exhaust holes

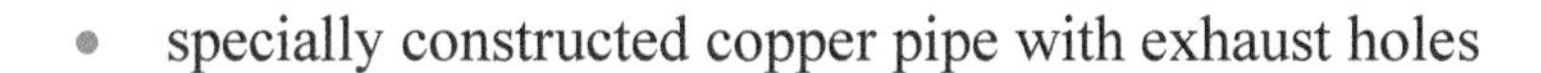

- compressed air or nitrogen gas
- safety glasses
- electrical heating tapes (optional)
- colored spotlights (optional)

PROCEDURE

You can produce an impressive cloud by the rapid boiling of liquid nitrogen. A suitable apparatus consists of a large (25-liter) Dewar flask filled with liquid nitrogen. A two-hole rubber stopper plugs the mouth of the Dewar flask. In one hole of the stopper place a tube connected to high-pressure air or through a pressure regulator to a cylinder of compressed nitrogen. In the other hole, place a tube that leads to the center of a meter-long, thick-walled, horizontal copper pipe, sealed at the ends, with several dozen small (1/8-inch-diameter) holes in the top through which the nitrogen can escape into the room. Wrap electrical heating tapes around the pipe to warm it prior to use. If a pipe with sufficiently thick walls is unavailable, insert a solid cylinder of copper loosely inside the pipe. The aim is to provide a warm orifice of high heat capacity to cause rapid boiling of the nitrogen for as long as possible. Depending on the heat capacity of the pipe and the flow rate of gas into the Dewar flask, you can produce a cloud reaching from the floor up several meters into the air and sustain it for a good fraction of a minute until the pipe cools to too low a temperature. You cannot repeat the demonstration for about an hour because the pipe has to warm back up to room temperature.

The operation of the device involves forcing the liquid nitrogen from the Dewar flask up into the warm pipe, where it rapidly boils and exits from the holes in the pipe as cold nitrogen gas. The cold gas then condenses water vapor from the air in the room above the apparatus and forms the cloud. Thus the demonstration works best when the humidity is high. If there is a ventilation system in the room, turn it off a few minutes before the demonstration to allow the humidity to rise. After about 20 seconds, the pipe will cool to the point where fountains of liquid emanate from the holes. The liquid is not nearly as effective as the cold gas in cooling the air, and the cloud slowly subsides. Colored lights illuminating the cloud provide extra visual appeal but may not provide adequate illumination for videotaping. The air is noticeably cooler after the cloud has dissipated especially near the floor, and children like to run down and play in the cloud. The demonstration is an effective conclusion to a presentation, and you can disappear in the cloud and perhaps exit the room through a nearby door unseen by the audience or walk through the cloud from behind to take a bow.

DISCUSSION

Although this demonstration mostly provides drama, it also illustrates a number of physical principles including Pascal's law, heat of vaporization, heat capacity, heat transport, and the condensation of water vapor in air as the temperature drops. The mechanism is similar to the way clouds in the sky form by cooling of the air to a temperature below which the air becomes saturated with moisture (100% relative humidity) [1−5]. The cloud consists of extremely small droplets of liquid water (about a thousandth of an inch in diameter) and is not steam or smoke as many people will respond if asked.

The water droplets in a cloud fall at a typical rate of about 0.5 cm/s, but real clouds do not fall from the sky because small updrafts of the same order support them and because the droplets drift slowly out the bottom of the cloud and evaporate while being replaced by new droplets forming above. The terminal speed of a small water droplet (or other sphere) is proportional to the square root of its radius.

A mixture of substances that may be in different phases (solid, liquid, or gas), such as a cloud (a liquid suspended in a gas), is called a "colloid." Colloids are common in nature, as a little thought will reveal. Carbonated beverages contain carbon dioxide gas (bubbles) immersed in a liquid. Smoke and dust are solids suspended in a gas. Mud consists of small solid particles of dirt immersed in a liquid as does paint and ink. A marshmallow is a gas suspended in a solid. Asphalt is a liquid suspended in a solid. The term "colloid" (which means "glue" in Greek) was coined by Thomas Graham (1805 – 1869) in 1861. In 1903 Wolfgang Ostwald (1883 – 1943) formulated the official definition of a colloid as "a system containing entities having at least one length scale in between 1 nm and 1 μm." At smaller sizes the system is considered a solution, and at larger sizes the particles fall under the influence of gravity. The study of colloids is known as "mesoscopic" physics (intermediate between microscopic and macroscopic).

HAZARDS

A possible hazard is explosion of the Dewar flask due to over-pressurization. The exit holes should be unobstructed, and you should not use pressures higher than about 30 psi unless you know that the Dewar flask is of adequate strength. It is wise to test the Dewar flask behind a suitable barricade at a pressure about 50% greater than what you intend to use. Some liquid nitrogen will exit from the holes, especially after the device has run for a while, and thus to avoid frostbite, do not allow your face and other body parts to be directly above the holes in the pipe when it is turned on. Wear safety glasses when working with liquid nitrogen. Before you activate the device, the pipe may be sufficiently hot to cause burns if heating tapes are used. The heating tapes also pose a potential electrocution hazard if nicked by something sharp thereby exposing the electrical conductors. If done in too small a room, there is the remote possibility of the nitrogen expelling enough oxygen from the room to suffocate someone, but this would be an extreme situation.

REFERENCES

1. N. H. Fletcher, *The Physics of Rain Clouds*, Cambridge University Press: Cambridge (1962).

2. C. F. Bohren, *Clouds in a Glass of Beer: Simple Experiments in Atmospheric Physics*, John Wiley & Sons: New York (1987).

3. R. R. Rogers and M. K. Yau, *A Short Course in Cloud Physics*, Pergamon: Oxford (1994).

4. C. F. Bohren and B. A. Albrecht, *Atmospheric Thermodynamics*, Oxford University Press: Oxford (1998).

5. M. T. Graham, *Phys. Teach.* **42**, 301 (2004).

2.13 Heat Convection

A candle extinguishes when a tightly fitting glass cylinder is placed over it unless a T-shaped piece of metal is lowered into the cylinder, illustrating natural convection.

MATERIALS

- candle
- glass tube, 30 cm long × 3 cm diameter
- metal tee
- matches
- incense sticks or a cotton swab with titanium tetrachloride (optional)
- coil of copper wire (optional)
- two candles and a wide-mouth jar with a removable lid (optional)

PROCEDURE

You can demonstrate the convection of heated air quite simply and effectively with a candle [1]. Light the candle to show the audience that it is quite ordinary. Hold your hand over the flame, and comment that the air rising from the candle is quite warm. Explain that this rising air pulls in air from below whose oxygen is necessary for combustion. Then lower a glass cylinder, about 30 cm long and 3 cm in diameter, open at both ends, over the candle, making a tight seal at the base so that air cannot enter from below. After a few seconds, the candle will extinguish. Then lower a thin, tightly fitting partition made of copper or some similar metal into the cylinder to within a few centimeters of the top of the candle. A T-shaped extension at its top holds the partition in place. Lift the cylinder, relight the candle, and put the cylinder back over the candle. The candle continues to burn with a flickering flame while the cylinder is in place. Finally, after a while, remove the partition, and let the candle go out. For each case, ask the audience to predict whether the candle will continue to burn. With a bit of practice, you can let the candle apparently go out and then revive it by quickly lowering the partition. Use an incense stick or a

cotton swab dipped into titanium tetrachloride to make smoke, which runs down one side of the partition and up the other side to make the convection more visible.

As an extension of the demonstration, extinguish the candle by abruptly touching a coil of copper wire to the flame [2]. The copper conducts the heat away and lowers the temperature below the kindling point of the candle wax.

Another related demonstration you can try is to put two burning candles, one taller than the other, in a wide-mouth jar and ask the audience which candle will extinguish first if you cover the top of the jar. You might expect the lower candle to extinguish first since the carbon dioxide (CO_2) produced by combustion has a higher molecular weight than the oxygen (O_2) that is consumed (44 versus 32), but in fact the opposite occurs. The reason is that the carbon dioxide is hotter than the air and thus has a lower density and rises to the top of the jar [3].

DISCUSSION

In the absence of the partition, the upward rising warm air interferes with the downward falling cool air near the top of the cylinder because of its viscosity, and the convection pattern does not reach down to the level of the candle. With the partition in place, the warm air rises up one side of the partition, and the cool air falls down the other side and replenishes the candle with oxygen. Whether the convection goes clockwise or counterclockwise around the partition depends on small asymmetries when the convection starts. Once the convection starts in one direction, it will tend to continue in that direction as long as the candle burns, much as the direction of the swirling water leaving the drain in a bathtub depends mainly on the small vorticity present in the water when the drain is opened [4].[1]

Along with radiation (see section 2.14) and conduction, convection is one of the ways in which heat transports from one place to another. Convection can occur in gases and liquids. Pressure gradients drive the convection in the fluid. With the candle, the pressure gradient is small, and so the warmer air is less dense than the colder air, and thus gravity does not attract it as strongly. The burning of a candle requires gravity, as you can demonstrate by observing that a candle in free-fall will extinguish [5]. A candle would not continue to burn in the weightlessness of outer space [6, 7]. Such experiments were actually carried out on the Space Shuttle, but inside a sealed chamber for safety reasons. One candle burned for about two minutes and another for about 20 seconds before going

---

[1] The Coriolis force due to the rotation of the Earth will determine the direction of the swirl if the initial vorticity is sufficiently small, which is rarely the case for laboratory-sized fluids. This force is in opposite directions in the Northern and Southern Hemispheres and dictates the rotation direction of tornadoes and hurricanes.

out. The flames were weak, spherical, and pure blue. With even a small flow of air, the candle burns continuously, and with elevated concentrations of oxygen (35–50%), it burns continuously even without the flow.

This demonstration illustrates natural convection, in which the thermal gradient is the driving mechanism. Heat can also be transported by forced convection in which a fan or blower is used to transmit air at one temperature to a location that is either hotter or cooler than the air. Buildings are usually heated by forced convection.

HAZARDS

The hazards in this demonstration are rather apparent and include burns from the candle and cuts from the glass, especially if it breaks. The partition will get quite warm at the bottom, and so you should take care to touch it only at the top.

REFERENCES

1. J. S. Miller, *Physics Fun and Demonstrations*, Central Scientific Company: Chicago (1974).
2. T. L. Liem, *Invitations to Science Inquiry*, Ginn Press: Lexington, MA (1981).
3. P. Hopkinson, *Phys. Teach.* **39**, 12 (2001).
4. J. C. Salzsieder, *Phys. Teach.* **32**, 107 (1994).
5. B. Z. Shakhashiri, *Chemical Demonstrations*, The University of Wisconsin Press: Madison, WI, Vol 2 (1985).
6. J. McKinley, *Phys. Teach.* **20**, 261 (1982).
7. H. F. Meiners, *Physics Demonstration Experiments*, Vol I, The Ronald Press Company: New York (1970).

2.14 Heat Transmitter

A match at the focal point of a parabolic reflector ignites by the radiation from an electrical heating element placed at the focal point of a second parabolic reflector across the room and aimed at the first.

MATERIALS

- two parabolic reflectors[1]
- electrical heating element
- wooden match
- piece of paper (for alignment)
- gunpowder (optional)
- plate of glass (optional)
- TV remote control, silicon solar cell, amplifier, and loudspeaker (optional)
- cotton swab dipped in liquid nitrogen and thermometer (optional)

PROCEDURE

You can effectively demonstrate the transfer of heat by radiation with two parabolic reflectors, preferably having a diameter of 20 centimeters or more. Place an electrical resistive heating element near the focus of one reflector. Place a wooden match at the focus of the other reflector. Then place the reflectors several meters apart and aimed at one another. When you energize the heating element, the match should burst into flame

---

[1] Available from American 3B Scientific, Fisher Science Education, Frey Scientific, Sargent-Welch, and Science Kits & Boreal Laboratories.

within about a minute. You can show that it is safe to stand in the beam since the power flux (watts per square meter) is quite low. Comment that you have never caught on fire doing this before. You can also place a plate of glass between the mirrors and point out that the radiation passes through the glass.

Mention that this is a more intense form of the same kind of radiation used in TV remote controls. With a silicon solar cell connected to an amplifier and loudspeaker, you can hear the sounds corresponding to the infrared signal from a remote control [1].

In a variation of the demonstration, replace the match with a cotton swab that you have dipped into liquid nitrogen to cool it, and place a small thermometer at the focal point of the opposite mirror to show that cold (the absence of heat) can also be focused and transmitted.

The reflectors need not be of optical quality since the heat source is quite spread out, but the surfaces should reflect infrared radiation with high efficiency. Chrome plating of the reflectors is helpful and attractive. A less effective variation uses a single reflector to focus an image of the lamp on the match [2]. You can use a piece of paper in a dimly lit room to view the red spot on which the heat is focused, enabling you to place the match in the optimal location. Avoid bumping either of the reflectors after you have aligned them. For a bit more drama, replace the match with a small amount of gunpowder, as Sir Humphry Davy (1778 – 1829) was fond of doing in his public lectures at the Royal Institution in London [3].

DISCUSSION

Radiation is one of the three ways in which heat transfers. The others are conduction and convection (see section 2.13). Only radiation can take place in a vacuum. The other mechanisms require a material medium. This is the means by which the Sun heats the Earth. Heat radiation is a form of electromagnetic wave in the infrared portion of the spectrum. It is of the same form as light, microwaves, radio waves, X-rays, and so forth, except of a different wavelength. When the temperature of a body increases, it emits successively shorter wavelengths, first giving off heat and then light. Most materials melt before they can give off significant amounts of ultraviolet radiation. The amount of energy radiated by a body is proportional to the fourth power of its absolute temperature (Stefan's law). Infrared radiation obeys the same laws of geometrical optics as does light, and thus you can focus it with mirrors and lenses as in this demonstration. You can also explain radiation in terms of the emission and absorption of individual photons.

HAZARDS

The heating element can cause electrocution as well as burns if operated directly from the power lines. A bad burn can result if any part of the body or clothing remains in the focus of the receiving reflector. Do not try to align the reflectors by placing your eye at the focus; rather, observe the spot on a piece of paper. Provide a means to extinguish the match safely once you light it and avoid dropping the match onto something that might catch fire. If you use gunpowder, experiment with small amounts of it to get an explosion that is impressive but safe.

REFERENCES

1. H. Manos, *Phys. Teach.* **35**, 552 (1997).
2. R. E. Berg, *Phys. Teach.* **28**, 56 (1990).
3. J. Tyndall, *Heat Considered as a Mode of Motion* (6th ed.), Longmans, Green, and Company: London (1880).

2.15 Kinetic Theory Simulator

A collection of small ball bearings in an enclosure, with one side connected to a loudspeaker, exhibits random motion analogous to the molecules in a gas.

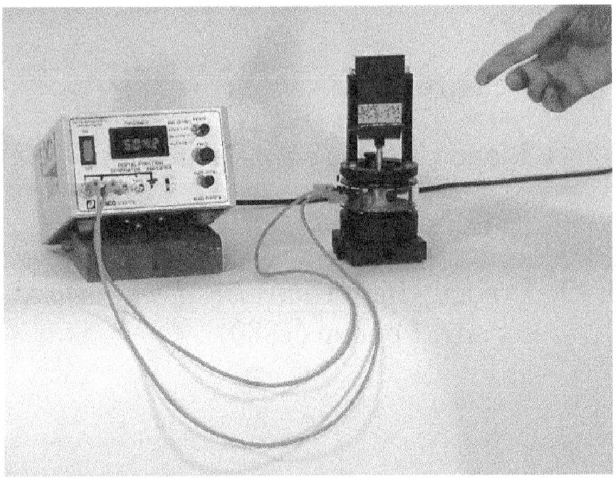

MATERIALS

- kinetic theory simulator[1]
- audio oscillator
- slide projector, arc lamp, overhead projector, or video camera and monitor (optional)
- matches (optional)
- drill press with wooden dowel and block of wood (optional)
- shallow glass of water with pepper on overhead projector (optional)

PROCEDURE

Place a dozen or more small plastic or steel balls in a transparent container, one boundary of which is made to oscillate rapidly back and forth by a vibrator or loudspeaker connected to an audio oscillator at a frequency of a few Hertz. You can make a larger version with Ping-Pong® balls in a large Plexiglas® cylinder with a motor-driven piston. For a large audience, place the apparatus on an overhead projector, form a shadow with a slide projector or an arc lamp against a screen, or view it with a video camera and monitor. You can vary the amplitude or frequency of the driving oscillation to illustrate the effect of changing the temperature of a gas. A lid resting on the balls but free to move upward illustrates the pressure exerted by molecules in motion as they bombard the walls of their container. You should emphasize that this demonstration is merely a simulation since typical molecules in a gas are about ten million times smaller in diameter than the balls used in the demonstration.

You can illustrate the relation of motion to heat by having the audience members rub their hands rapidly together and then place them against their cheeks. The hand rubbing is an ordered motion, but the energy of that motion changes into the disordered motion of the molecules in their hands through friction. There are many other demonstrations

---

[1] Commercial versions, sometimes called "molecular motion demonstrators" are available from Carolina Biological Supply Company, Frey Scientific, PASCO Scientific, and Sargent-Welch.

showing the conversion of ordered motion into heat. Striking a match is a dramatic example. Try placing a wooden dowel in a drill press and try to use it to drill into a second piece of wood. Point out that if you were ever lost in the woods, you could make a fire this way if you had a drill press handy.

Put a shallow glass of water with some pepper sprinkled on its surface on an overhead projector to observe the Brownian motion of the pepper. Brownian motion was discovered by Robert Brown (1773 – 1858), a Scottish biologist, who in 1827 noticed the incessant motion of pollen grains suspended in water when viewed under a microscope. He demonstrated that the motion was physical rather than biological by boiling and freezing the water to kill any living organisms.

A good audience participation demonstration appropriate for children is to have a group of them pretend they are molecules and to move in slow motion when they are cold and rapidly when they are hot. Demonstrate the motion, and tell them it is all right to collide occasionally with one another. Point out that molecules are not very smart, and they just move about aimlessly until they collide with another molecule or with the walls of their container.

DISCUSSION

Simulations such as this provide a useful means for visualizing the motion of molecules [1, 2]. An ideal gas obeys the ideal gas law

$$pV = NkT$$

where p is the pressure, V is the volume, N is the number of molecules in the volume, k is Boltzmann's constant (1.38×10^{-23} J/K), and T is the temperature (in kelvins). Thus with a given number of molecules N in a constant volume V, the pressure increases in proportion to the temperature. This is known as Charles's Law after Jacques Alexandre Cesar Charles (1746 – 1823), who proposed it in 1787 but never published it, but it is also known as Gay-Lussac's law after Joseph Louis Gay-Lussac (1778 – 1850), who verified and published it in 1802 [3]. If the volume V can change but the pressure p remains constant because of a movable lid of given weight, the volume increases in proportion to the temperature.

Real gases would typically be in thermal equilibrium with the walls of their container, and thus the molecules would on average gain as much energy as they lose upon collision with the walls and with one another. In the demonstration here, the kinetic energy of the balls is considerably greater than their thermal energy at room temperature (about kT), and thus the balls lose energy every time they collide with the walls. Hence it is necessary to provide energy input for the balls to continue moving. You can calculate the effective "temperature" of the simulated gas from

$$kT = mv^2$$

where m is the mass of the balls and v is their velocity. If the balls are 0.01 kg and their velocity is 10 m/s, the temperature is about 10^{23} degrees!

This demonstration also illustrates how apparent randomness can arise from simple equations of motion if there are many particles interacting. Even without gravity and with perfectly elastic collisions (no energy loss), the motion is very complicated and effectively unpredictable after several collisions. If you paint one of the balls red and follow its motion through many collisions, you will observe that it executes a random walk or Brownian motion (see section 1.21).

HAZARDS

Take care not to burn out the loudspeaker with too large of an input voltage. A handful of ball bearings carelessly spilled on the floor can make for treacherous walking!

REFERENCES

1. B. Z. Shakhashiri, *Chemical Demonstrations*, The University of Wisconsin Press: Madison, WI, Vol 2 (1985).

2. S. Mak and D. Cheung, *Phys. Teach.* **39**, 48 (2001).

3. J. Guy-Lussac, *Annales de Chimie* **43**, 137 (1802).

2.16
Carbon Dioxide Trough

Carbon dioxide from a glass beaker pours down a trough containing a number of candles that successively extinguish when contacting the invisible gas.

MATERIALS

- V-shaped trough with a number of candles mounted inside[1]
- 2-liter glass beaker
- cardboard cover for beaker
- small block of dry ice or vinegar and bicarbonate of soda
- tongs or gloves
- matches
- safety glasses
- hot plate (optional)
- pan of hot water (optional)
- balloon (optional)

PROCEDURE

A V-shaped Plexiglas® trough inclined at an angle of about 30° contains about five candles mounted vertically within the trough with their wicks below the top of the trough. Light the candles in honor of someone's birthday. "Happy Birthday" music is a nice touch here. There is a better than even chance that someone in the audience will have a birthday on any given day if there are at least 250 people in the audience (see table 2.3). If no one's birthday is today, ask about yesterday or tomorrow or last week.

Remove a 2-liter glass beaker containing carbon dioxide gas from behind the lecture bench and ask the audience what is in the apparently empty beaker. Then pour its

[1] Available from Frey Scientific.

contents down the trough. The candles go out one by one as if there were water pouring down the trough, but the beaker is apparently empty since carbon dioxide is a colorless, odorless gas. Explain that the beaker contained carbon dioxide and that fire extinguishers often use carbon dioxide because it does not support combustion.

You can produce the carbon dioxide by placing a piece of dry ice in the beaker before the lecture with a piece of cardboard on the top to cover the beaker. Remove the cardboard, and surreptitiously lift the dry ice out with tongs or gloves just before you show the beaker to the audience. You can also put the dry ice in a smaller beaker inside the large beaker for easier removal. You can store the beaker on a slightly warm, electric hot plate to prevent the formation of frost on its exterior and to ensure the rapid evaporation of the dry ice. You can often procure dry ice from ice cream vendors for about $1 per pound and can store it for about a day in a Dewar flask, thermos bottle, or cardboard box, preferably in a refrigerator or freezer.

Ask the audience why it is called "dry" ice, and explain that it does not make things wet because it changes directly from a solid into a gas without ever becoming a liquid. Ask if anyone knows what that process (sublimation) is called. Ask whether water ice can sublimate, and explain that snow will slowly evaporate even if the temperature does not rise above freezing and that ice cubes in the freezer eventually evaporate.

You can dump the dry ice into a pan of hot water to make a cloud when you are done. Explain that the cloud is not carbon dioxide, but tiny droplets of liquid water that condense from the air when it cools by contacting the cold carbon dioxide gas. Point out that our automobiles emit more than their own weight in carbon dioxide every year [1].

Another way to conclude the demonstration is to place a bit of dry ice inside a balloon and tie it off. The balloon will expand to a large size as the dry ice sublimates and fills the balloon with carbon dioxide vapor, eventually popping the balloon if you use enough dry ice.

If dry ice is not available, you can make carbon dioxide by mixing a small amount of vinegar (dilute acetic acid or CH_3COOH) with a small amount of baking soda (sodium bicarbonate or $NaHCO_3$), although this method leaves a visible residue in the beaker. The reaction is $CH_3COOH + NaHCO_3 \rightarrow CH_3COONa + H_2O + CO_2$ (gas).

DISCUSSION

This demonstration illustrates the necessity of oxygen for sustaining combustion. It also demonstrates sublimation, the process by which a liquid changes directly from a solid to a gas without going through the liquid phase. Dry ice is called "dry" because, unlike ordinary water ice, it sublimates at atmospheric pressure. Ordinary ice sublimates (and hence is also dry) when the vapor pressure of the water above the liquid is below 4.58 Torr (or 0.006 atmospheres). This condition can exist on a dry day and is the reason snow and ice eventually disappear, even when the temperature is below freezing (0°C), and the reason ice cubes stored in the freezer slowly vanish without leaving a puddle of liquid water. The reverse process also occurs, such as when water vapor condenses into a solid to make snow, slowly growing a crystal with a unique and beautiful shape. The carbon dioxide remains in the beaker because it is about 50% more dense than air. Because it is colder than the surrounding air, it is even denser than it would be at room temperature.

Dry ice is at a temperature of −78.5°C or −109.5°F. Liquid carbon dioxide cannot exist below the triple point pressure of 5.1 atmospheres and temperature of −56.2°C. You can observe liquid carbon dioxide by letting dry ice sublimate in a heavy-walled transparent container, thereby raising its pressure above the triple point [2].

Some types of fire extinguishers use carbon dioxide. When the fire extinguisher is used, the liquid boils vigorously, then cools into a fine snow, and then sublimates back into a gas. The atmosphere of the Earth contains about 0.035% carbon dioxide, but carbon dioxide constitutes about 95% of the (thin) atmosphere of Mars (less than 1% of Earth's atmospheric pressure) and about 96% of the (thick) atmosphere of Venus (about 100 times Earth's atmospheric pressure), causing their temperatures to be even higher (greater than 800°F for Venus) than would otherwise be expected. Too much carbon dioxide in the Earth's atmosphere from burning fossil fuels is producing troublesome global warming [3–5].

HAZARDS

Dry ice sublimates at a temperature of −78.5°C and can cause frostbite. Thus you should handle it with tongs or insulated gloves and not allow it to touch bare skin. Do not seal dry ice in a container because it can cause the container to explode as it sublimates. A Styrofoam® ice chest with a loosely fitting lid makes a good container for transporting dry ice. When crushing or grinding dry ice, wear safety glasses to prevent getting any of the dust into your eyes.

REFERENCES

1. C. Sagan, *Billions and Billions*, Random House: New York (1997).

2. T. C. Lamborn and R. Mentzer, *Phys. Teach.* **32**, 508 (1994).

3. J. K. Leggett, *The Carbon War: Global Warming and the End of the Oil Era*, Routledge: New York (2001).

4. S. R. Weart, *The Discovery of Global Warming*, Harvard University Press, Cambridge, MA (2003).

5. J. T. Houghton, *Global Warming: The Complete Briefing* (3rd ed.), Cambridge University Press: Cambridge (2004).

Table 2.3

Probability that someone in an audience of a given size will have a birthday on a given day

Audience Size	Probability
10	3%
20	5%
30	8%
70	13%
100	24%
150	34%
200	42%
250	50%
300	56%
400	67%
500	75%

2.17 Weight of Air

A hollow sphere and a balance scale with a vacuum pump demonstrate that air has weight.

MATERIALS

- balance scale
- small weights
- hollow sphere
- vacuum pump
- basketball (optional)
- bell jar (optional)

PROCEDURE

There are at least two ways to demonstrate that air has weight. The most direct demonstration is to place a hollow sphere on a balance scale and to bring it to a balance with a number of small weights [1]. Then attach the sphere to a vacuum pump, which removes the air from the sphere, causing the balance to change. Alternately, evacuate the sphere first and bring it to a balance. Then let air into the sphere, causing it to become unbalanced. You can add weights to the other side of the balance scale until you restore the balance and then note how much weight you added to see what the air in the sphere weighs. If you do not have a sphere that you can evacuate, you can measure the weight of a basketball before and after it is inflated [2] or use a cardboard box instead, into which you flow helium or other low-density gas through a hole in the bottom, allowing it to expel the air through other holes in the bottom of the box [3, 4].

The second version of the demonstration is slightly less direct and requires that you use a bell jar, but it does not require that you evacuate the sphere. Place the balance scale inside the bell jar with the sphere on one side balanced by weights on the other side of the balance. Then evacuate the bell jar, reducing the buoyancy on the sphere. The scale will become unbalanced. Let air back into the bell jar to restore the balance. Point out that air does not weigh very much, but that the weight is sufficient to permit demonstrations such as this. This version of the demonstration leads naturally into a discussion of buoyancy, or it could follow such a discussion to illustrate the effect.

DISCUSSION

A sphere with a diameter of 3 inches (7.62 cm) has a volume of 232 ml, corresponding to about 0.01 moles since 1 mole occupies 22.4 liters at standard temperature and pressure (0°C and 760 Torr). Since air has a molecular weight of about 29 g/mole, the air in the sphere weighs about 300 milligrams. In the first version of the demonstration, you observe this weight directly. In the second version, you infer it indirectly through the reduced buoyancy of the sphere when you remove the surrounding air. The buoyancy of an object is equal to the weight of the fluid displaced by the object (Archimedes' principle).

HAZARDS

The main hazard with this demonstration is implosion of the bell jar, which has several tons of atmospheric pressure on it while evacuated. Use only bell jars designed to be evacuated, and be careful not to drop the bell jar or strike it with any hard objects. Make sure the belt of the vacuum pump has a guard, or place the pump out of reach.

REFERENCES

1. R. G. Buschauer, *Phys. Teach.* **29**, 115 (1991).

2. H. Brody, *Phys. Teach.* **27**, 46 (1989).

3. J. H. Pepper, *The Boy's Playbook of Science*, Routledge, Warne, and Routledge: London (1860).

4. T. B. Greenslade, *Phys. Teach.* **31**, 160 (1993).

2.18
The Impossible Balloon

A specially constructed balloon appears to have a lifting power far beyond that permitted by Archimedes' principle.

MATERIALS

- specially constructed balloon
- weights
- ordinary balloon filled with helium with a string and weight (optional)

PROCEDURE

The late Professor Ed Miller (1915 – 1995) of the University of Wisconsin liked to perform a demonstration involving a buoyancy hoax in which a half-meter-diameter balloon is against the ceiling with a weight hanging below it. After explaining all about buoyancy, Professor Miller would go into an enthusiastic story about the exciting development by some of his colleagues of a new gas with 47 times the lifting power of hydrogen and how we are fortunate to get an advance demonstration. Then he would hang several more, obviously heavy weights to the string attached to the balloon. Eventually someone would wake up and argue that this is inconsistent with all he had been saying. He did not say a word, but just turned around, reached up, and pulled out the cork to release the gas from the balloon. The balloon expels its gas and collapses, but the weights stay hanging there. Obviously there is a string through the balloon to a small hook in the ceiling that holds the weights up. You can pass a string through a balloon on the end of a large needle that punctures the balloon at a place where you place some transparent Scotch® tape to prevent it from tearing. The balloon will slowly deflate over the course of about half an hour. Finally, a ring on top of the balloon, that kept it steady, falls down around the collapsed balloon onto the weights with a clatter, providing a humorous climax to the hoax.

In a simpler version of the demonstration, use a cluster of three or more helium-filled balloons attached to the string, which runs up between them to a hook on the ceiling. You can use a second cluster of similar helium-filled balloons on a short string to show how much (little) weight they can lift.

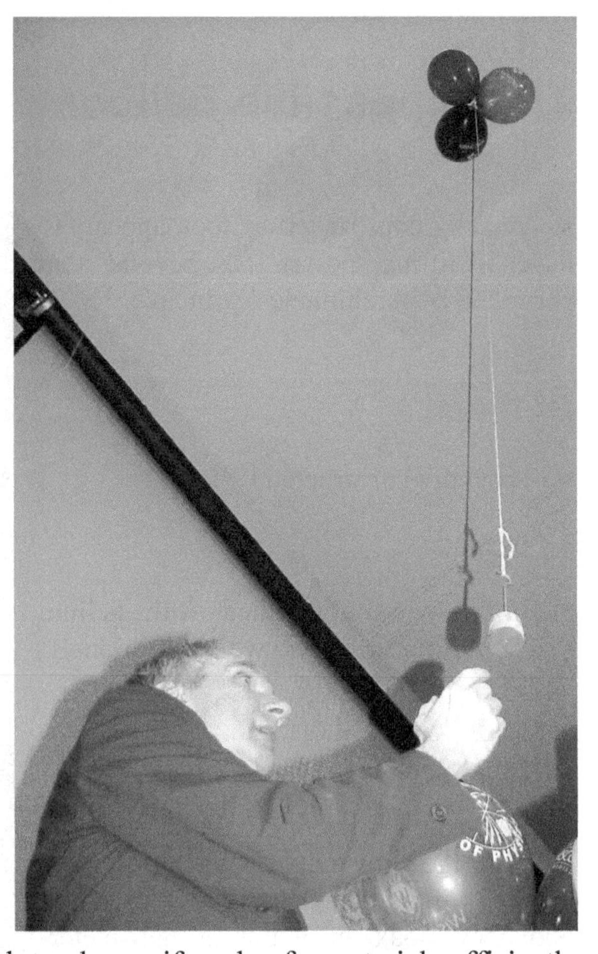

DISCUSSION

The amount of weight that a balloon can lift is equal to the weight of the air displaced by the balloon minus the weight of the balloon material and the gas that it contains. Thus hydrogen (H_2), with a molecular weight of 2 grams per mole compared to air with an average molecular weight of 29 grams per mole, is close to the best that one can do [1]. Helium (He) has a molecular weight of 4 grams per mole. Even a gas with zero molecular weight (a vacuum) would provide a lift only 7% greater than hydrogen. The use of an evacuated balloon is, of course, impractical since the balloon would be much too heavy if made of a material sufficiently strong to support the pressure of the atmosphere. The pressure in a hydrogen-filled balloon is slightly greater than the pressure of the surrounding atmosphere (typically a few percent), and so such a balloon would contain more moles of hydrogen than the air that is displaced, which increases its weight slightly. Since the upward buoyant force on a balloon is equal to the weight of the air that it displaces (Archimedes' principle), it depends only on the balloon's size and not on its contents.

HAZARDS

The only hazard in this demonstration is the string breaking and dropping the weights on someone's foot.

REFERENCE

1. A. W. Burgstahler, T. Wandless, and C. E. Bricker, *Phys. Teach.* **25**, 434 (1987).

2.19
Neutral-buoyancy Balloon

A helium-filled balloon attached to a heavy string rises until its buoyancy just balances its weight plus the weight of the string.

MATERIALS

- balloon
- helium gas
- ball of heavy string
- scissors
- Mylar® balloon and hair dryer (optional)

PROCEDURE

Inflate a balloon with helium and tie it to the loose end of a ball of heavy string either before or during the lecture. Holding the string close to the balloon, so that the balloon rises, ask the audience what gas is in the balloon. Most people will say "helium," but point out that it could be any gas less dense than air. Explain that the gas in fact is helium, and ask what will happen if you release the string. Most people will say it will rise to the ceiling. Then release it and let it rise until the string becomes taught. Then unwind a bit of string off the ball, and ask why it stopped rising. Most people will say it is because of the weight of the string. Explain that, indeed, the buoyancy of the balloon just balances the weight of the balloon and attached string. Then explain that by cutting the string at the point where it just touches the table or floor, you will have a "neutral-buoyancy balloon." Cut the string and release it with the lower end of the string at head level. If the balloon begins to fall, cut a bit more string off the end. If it slowly rises, point out that you did that on purpose because the helium slowly leaks out of the balloon, and that when this happens the balloon will descend. Suggest to the audience that when this happens they can take a fingernail clipper and cut a centimeter of string off the end and release it again.

Explain that Mylar® balloons hold their helium much better than rubber (typically latex) balloons, sometimes for weeks. Suggest that if someone ever gets such a balloon, it would be fun to attach a heavy string of the right length to it and release it inside the house. In this way, you can see where the air currents are flowing. The balloon slowly wanders around the house like a curious pet. Suggest that it would be a good wedding present for newlyweds on their wedding night. Meteorologists use weather balloons to measure the direction and strength of winds above the surface of the Earth. Point out that there are "winds" indoors as well as outdoors.

For a variation of the demonstration, use a Mylar® balloon partially filled with helium so that it is not fully inflated. Adjust the attached weight so that the balloon does not quite support the weight. Heat the balloon with a hair dryer so that it expands slightly and rises. As it cools, it will fall again. You can do the same demonstration with a rubber balloon, but it will not expand as much, and the effect will be less dramatic.

DISCUSSION

The buoyancy of a balloon is equal to the weight of the air displaced by the balloon. A spherical balloon with a diameter of 12 inches (30.5 cm) has a volume of 14.8 liters, corresponding to 0.66 moles since one mole occupies 22.4 liters at standard temperature and pressure (0°C and 760 Torr). Since air has a molecular weight of about 29 g/mole, such a balloon can lift a mass of about 19 grams. Thus the balloon, the helium enclosed, and the string must have a total mass of about 19 grams to achieve a state of neutral buoyancy. From these numbers, it is easy to estimate how large a balloon is required to lift a person or other object if you remember that the volume, and hence the buoyancy, increases with the cube of the diameter. If you pop the balloon and weigh it, you should get the corresponding value except for a small error from neglecting the mass of the helium that escapes.

Heating the balloon does not change its weight or the weight of the enclosed gas, but it does increase the volume of the balloon as a result of the increased pressure of the hot gas and hence the amount of air displaced by it, thus increasing its buoyancy. In a hot-air balloon, the pressure is nearly the same inside and outside the balloon, but the density is less for the enclosed hot gas than for the cold gas outside at the same pressure. Hot air balloons that carry people have nearly neutral buoyancy for obvious reasons. Most fish have gas bladders that can expand or contract to achieve a state of neutral buoyancy in the water. Astronauts practice maneuvering in weightlessness under water wearing diving suits with weights attached to make them neutrally buoyant.

If you release a balloon with positive buoyancy outdoors, it will expand as it rises due to the decreasing external pressure on it. The expansion would increase the buoyancy, except for the fact that the air that it displaces has a density that decreases with altitude. Thus the balloon will either burst if the rubber stretches too much or it will reach an altitude where its buoyancy just balances its weight [1].

On July 2, 1982, a truck driver named Larry Walters ("Lawnchair Larry") attached 42 six-foot-diameter helium-filled weather balloons to a Sears® lawn chair (dubbed "Inspiration I") in the backyard of his girlfriend's house in San Pedro, California. He ascended to an altitude of 16,000 feet over California, where he drifted for 45 minutes until shooting some of the balloons with a pellet gun. He landed on a power line in Long Beach [2], causing a blackout, and was promptly arrested and fined $1,500 for violating various FAA regulations. Despite having survived his ordeal, Walters received honorable mention in the 1982 Darwin Award, an honor presented every year to an individual (or the remains thereof) who has done the most to remove undesirable elements from the human gene pool. On October 6, 1993, at the age of 44, Walters wandered into the forest and committed suicide by shooting himself in the heart.

HAZARDS

There are no significant hazards with this demonstration. The use of hydrogen is not recommended if there are hot overhead lights that could ignite the gas. Don't try to replicate Larry Walters' flight!

REFERENCES

1. M. P. Silverman, *Phys. Teach.* **36**, 288 (1998).
2. M. Oliver, *Los Angeles Times*, 24 November (1993), p. 16.

2.20 Exploding Balloons

Helium and hydrogen-filled balloons tethered by strings above the lecture bench burst when touched by a lighted match on the end of a stick.

MATERIALS

- two identical balloons
- twenty feet of thin string
- helium and hydrogen gas[1]
- wooden stick about 1 meter long with an alligator clip on one end
- stick matches or a candle

PROCEDURE

Fill two identical balloons, one with helium and the other with hydrogen. Attach the balloons to either end of the lecture bench with thin strings that allow them to float about ten feet above the floor. You should fill the balloons within about an hour of using them since the gas will gradually diffuse through the balloons, and they will lose their buoyancy. Alternately, an assistant can enter the room on cue with the balloons.

Ask the audience to guess what is in the balloons. Most people will assume they contain helium. Then ask the audience to predict what will happen when a lighted match or candle is touched to one of the balloons. Most people will say the balloon will burst. If someone suggests that the balloons contain hydrogen, ask how we might determine whether the gas is hydrogen or helium. Look horrified when someone suggests igniting them.

Light a match and clamp it with an alligator clip attached to the end of a wooden stick about a meter long or light a candle attached to the end of the stick. Touch the match or candle to the helium balloon, causing it to burst but without burning. Point out that helium was discovered first on the Sun in 1868, and hence the name, after Helios, the Greek god of the Sun. Ask the audience if they would like to see you repeat it with the other balloon. Caution them to cover their ears if loud noises bother them. Then repeat

---

[1] You can procure specialty gases and equipment from Matheson Tri-Gas, 166 Keystone Drive, Montgomeryville, PA 18936, 800-416-2505, http://mathesongas.com/.

the demonstration with the hydrogen balloon, causing a large explosion and ball of flame. For a more dramatic demonstration, use a vertical column of hydrogen-filled balloons.

When the audience regains its composure, you can explain the importance of taking nothing for granted in science. Helium is only one example of a gas that is less dense than air. In fact, hydrogen is lighter yet (about half as dense). The net upward force, however, is only slightly different since it is equal to the difference between the weight of the gas in the balloon and the weight of the air displaced by the balloon (the buoyant force). You can discuss Archimedes' (287 BC – 212 BC) principle, chemical combustion, or the scientific method.

Mention the Hindenburg disaster, in which thirty-six people died on May 6, 1937, in Lakehurst, New Jersey, and the reasons hydrogen is no longer used in dirigibles. Most of the injuries from the Hindenburg came from the burning fabric and fuel rather than the burning hydrogen, which tends to rise out of the way. Turbulence can also destroy helium-filled balloons, as happened to the Shenandoah on September 3, 1925, near Ava, Ohio. You can discuss the operation of hot-air balloons [1] and explain that hot air is less dense than cold air because pressure is the product of the particle density and the absolute temperature (in kelvins), and thus for a given pressure, the density is inversely proportional to the absolute temperature.

You can also do the demonstration with other gases such as air and oxygen if you suspend the balloons in some way such as with a ring stand. You can use a stoichiometric mixture (2:1) of hydrogen and oxygen if you take proper precautions to protect people's ears from the loud explosion that results [2]. If hydrogen gas is not available, you can produce it in a 250 ml Pyrex® Erlenmeyer flask to which 10 to 15 grams of mossy zinc and 65 ml of 3M hydrochloric acid has been added [3]. You can also increase the sound released in the explosion of the hydrogen balloon by putting some air in the balloon before filling it with hydrogen. Experiment with various mixtures to get an appropriately dramatic but not uncomfortably loud explosion. You can also ignite the balloon with the spark from a Tesla coil (see section 4.6) or with a high-power laser (see section 6.3).

DISCUSSION

The tendency of a balloon to rise when filled with a gas less dense than air illustrates Archimedes' principle. The balloon will rise if its weight plus the weight of the gas inside is less than the weight of the air displaced (the buoyant force). The pressure of a gas depends only on the number of molecules per unit volume and the temperature, and not on the mass of the molecules. At room temperature and atmospheric pressure, a cubic meter of air weighs almost three pounds. You can make an air-filled balloon or soap bubble float in a bath of carbon dioxide produced by dry ice in the bottom of a transparent container such as an aquarium.

The combustion of the gas in a balloon is an illustration of a chemical reaction, in this case hydrogen in the balloon reacting with oxygen in the surrounding air to form water with the release of 232 kJ per mole of water formed. The reaction ensues much more violently if you mix oxygen with the hydrogen than if it has to mix with oxygen from the air. This gives some indication of the rate with which gases diffuse when you suddenly remove the partition between them (the balloon).

HAZARDS

This demonstration looks more dangerous than it is. If the area around and especially above the balloon is clear of obstructions and if the balloon is ignited using a stick at least a meter long at arm's length, it is relatively safe. You should practice with small balloons to gain confidence. Remember that the volume of hydrogen is proportional to the cube of the diameter of the balloon, and thus the demonstration quickly scales up to quite dramatic proportions. Mixtures of hydrogen and oxygen require ear protection for both the demonstrator and the audience, and even with pure hydrogen, you should caution people to cover their ears "just in case."

REFERENCES

1. L. A. Bloomfield, *How Things Work: The Physics of Everyday Life*, John Wiley & Sons: New York (2001).
2. B. Z. Shakhashiri, *Chemical Demonstrations*, The University of Wisconsin Press: Madison, WI, Vol 1 (1983).
3. S. Isom and C. Lail, *The Science Teacher* **56**, 45 (March 1989).

2.21 Exploding Soap Bubbles

Soap bubbles blown with natural gas or hydrogen are ignited with a candle as they rise toward the ceiling.

MATERIALS

- soap bubble solution
- glass pipe
- natural gas (methane) or hydrogen[1]
- candle
- matches
- oxygen (optional)

PROCEDURE

You can make soap bubbles by filling a glass pipe with a small amount of soap solution (sold in toy stores). Most people have seen bubble pipes or had one when they were younger. You can make suitable solutions by adding a few drops of sodium or potassium oleate or liquid dish detergent to a beaker of warm, distilled or soft, pure water. A small amount of glycerin and a few drops of ammonia water greatly improve the lasting quality of the bubbles [1].

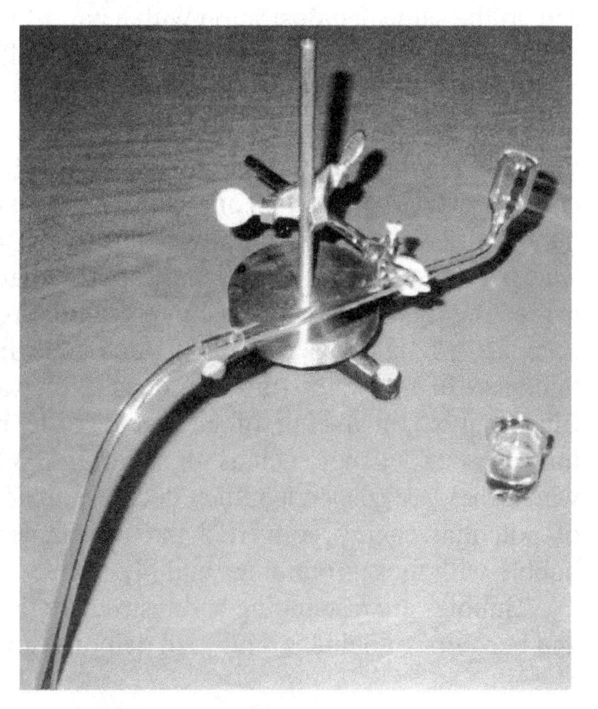

Support the pipe by a ring stand, and connect it to a source of compressed gas. A gas less dense than air, such as helium, allows the bubbles to rise to the ceiling as you release them from the pipe. Allow the gas to flow through the pipe briefly before

[1] You can procure specialty gases and equipment from Matheson Tri-Gas, 166 Keystone Drive, Montgomeryville, PA 18936, 800-416-2505, http://mathesongas.com/.

filling it with soap solution to expel the air. If you blow the bubbles with methane (natural gas) or hydrogen, you can ignite them with a candle held in your hand as they rise [2, 3]. You can mix some oxygen with the hydrogen to make a much louder explosion. The loudest explosion will occur when the hydrogen-to-oxygen ratio has a stoichiometric value of 2:1 so as to produce pure H_2O as the combustion product. Natural gas, often conveniently available in lecture rooms, makes a beautiful and quiet flame. Natural gas consists primarily of methane (CH_4), which is slightly less dense than air. Point out that this is the same kind of gas that many people have in their homes to run their furnace, hot water heater, and stove. Bubbles filled with mixtures of methane and oxygen will rise only if the ratio of CH_4 to O_2 is large since oxygen is slightly denser than air. The effect is best if viewed against a dark background in subdued illumination.

DISCUSSION

This demonstration illustrates a number of physical concepts. The very existence of bubbles is a demonstration of surface tension. A bubble consists of a thin film of liquid, held together by surface tension. It has elastic properties similar to those of a balloon. The pressure of the gas inside the bubble and the radius and thickness of the water film determine the size of the bubble. You can illustrate that the gas inside a bubble is under pressure by blowing a bubble with a pipe, but before you release the bubble, allow it to deflate and blow out a candle near the stem of the pipe. A smaller bubble has a higher pressure than a larger bubble for the same reason that a balloon is hard to blow up at first but becomes easier as it gets larger. You can illustrate this fact with a T-shaped pipe in which two bubbles of different size are blown and then connected together by means of the pipe. The larger bubble will get larger, and the smaller bubble will get smaller. You can do the same demonstration with a pair of unequally sized balloons.

The tendency of a bubble to rise when filled with a gas less dense than air illustrates Archimedes' principle. The bubble will rise if its weight plus the weight of the gas inside is less than the weight of the air displaced. At room temperature and atmospheric pressure, a cubic meter of air weighs almost three pounds. For the same reason, a helium balloon with a volume of one cubic meter can lift a weight of nearly three pounds. A bubble filled with air will slowly fall to the floor because of the weight of the film of water. The weight of the air inside approximately cancels the weight of the air displaced, neglecting the small difference in pressure. From the radius of the bubble and its terminal speed as it falls (see section 1.1), you can estimate the weight of the water and hence the thickness of the water film (the density of water is 1,000 kg/m^3). You can float an air-filled bubble in a bath of carbon dioxide produced by dry ice in the bottom of a transparent container such as an aquarium. When bubbles or balloons rise in the fluid in which they are suspended, they decrease their potential energy as they rise, with the loss of potential energy converted into kinetic energy and heat due to the friction of the bubble with the surrounding fluid [4].

Bubbles are fascinating and instructive. Entire lectures can be given on the properties and behavior of bubbles, and such demonstrations [5–9] inevitably enthrall the audience.

HAZARDS

Aside from the normal precautions involved with flames, the oxygen mixtures can make a very loud explosion. If you use a room with a low ceiling, be sure to ignite the bubbles quickly after they leave the pipe and not directly underneath something that might catch fire. You should use earplugs and caution the audience to hold their hands over their ears if you use the hydrogen and oxygen mixture.

REFERENCES

1. C. L. Strong, *Scientific American* **220**, 128 (May 1969).
2. B. Z. Shakhashiri, *Chemical Demonstrations*, The University of Wisconsin Press: Madison, WI, Vol 1 (1983).
3. H. A. Robinson, Ed., *Lecture Demonstrations in Physics*, American Institute of Physics: New York (1963).
4. D. Keeports, *Phys. Teach.* **40**, 164 (2002).
5. C. V. Boys, *Soap Bubbles and the Forces Which Mould Them*, Educational Services Incorporated, Doubleday: Garden City, NY (1959).
6. K. J. Mysels, K. Shinoda, and S. Frankel, *Soap Films*, Pergamon: New York (1959).
7. C. Isenberg, *The Science of Soap Films and Soap Bubbles*, Tiesto Ltd.: Clevedon (1978).
8. S. Simon, *Soap Bubble Magic*, Lathrop, Lee and Shepard: New York (1985).
9. A. Ward, *Experimenting with Surface Tension and Bubbles*, Dryad Press: London (1985).

2.22 Nonburning Handkerchief

A cotton handkerchief or dollar bill is immersed in a liquid and set on fire, but the handkerchief or bill does not burn.

MATERIALS

- cotton handkerchief or dollar bill
- 50% solution of isopropyl alcohol and water
- small amount of table salt (sodium chloride)
- matches or cigarette lighter
- tongs, at least 30 cm long
- bucket of water (recommended)
- coin and cigarette (optional)

PROCEDURE

Take a brightly colored, cotton handkerchief from your pocket, or better yet, from a "volunteer" planted in the audience and soak it in a liquid and set it on fire [1]. You can say that the handkerchief is dirty and that the solution is a cleaner. Pretend that it was accidentally set on fire while holding it over a candle to dry after cleaning it in the liquid. While holding the handkerchief with tongs and watching the flame, point out that the handkerchief is not burning. Finally snuff out the flame with a quick jerk of your wrist or douse it in a bucket of water and pass it around for inspection or return it to the volunteer. It might be hard to extinguish the flame with only a flick of the wrist, and the additional air might in fact make it burn brighter. For added drama, have an assistant put out the flame with a fire extinguisher. Ask the volunteer to hold up the handkerchief for everyone to see. Explain that this is a trick magicians often do.

You can also use other combustible materials, such as paper, in place of the handkerchief [2]. A dollar bill provided by someone in the audience is especially effective. Make a comment about always wanting to have enough money to burn or about physicists or teachers deserving more money.

In this demonstration, the liquid is a 50/50 mixture of isopropyl alcohol (the kind sold as "rubbing alcohol") and water. You can substitute methanol or ethanol for the isopropyl alcohol. Some table salt in the solution will give the flame a yellow color and will help make it more visible. Note carefully the purity of the alcohol before mixing. Rubbing alcohol as sold in drug stores is often 30% or more water, and the alcohol evaporates more quickly than the water, diluting the solution. You could repeat the demonstration with varying mixtures of alcohol and water. Too much water will prevent the alcohol from burning, and too little water will allow the cloth or paper to char.

In a variation of the demonstration, wrap a dry cotton cloth tightly around a coin, and touch a lighted cigarette to the cloth. The coin absorbs the heat and keeps the temperature below the point where the cloth burns. Comment on the dangers of smoking if you wish.

DISCUSSION

This demonstration illustrates the variation in the temperature required to support combustion in different substances. The alcohol burns at a temperature below the kindling temperature of the cotton or paper. In addition, the heating and vaporization of the water removes heat and prevents the cloth or paper from burning.

HAZARDS

Although the flame is relatively cool, it can cause severe burns and ignite other things that burn much hotter. Hold the handkerchief with a pair of long tongs (30 cm or more) while it is ignited. Plan for a way to extinguish the flame before the water completely vaporizes or the burning object falls to the floor. Alcohol can damage the eyes severely, and so you should wear safety glasses with this demonstration.

REFERENCES

1. B. Z. Shakhashiri, *Chemical Demonstrations*, The University of Wisconsin Press: Madison, WI, Vol 1 (1983).

2. J. Jardin, P. Murray, J. Tyszka, and J. Czarnecki, *Journ. of Chemical Education* **55**, 655 (1978).

2.23 Ethanol Vapor Explosion

A small amount of ethanol placed in a bottle is made to explode and blow a cork a considerable distance by means of an electrical spark.

MATERIALS

- 250-ml polyethylene bottle with two nails or screws through the sides
- cork (or rubber) stopper to fit the bottle
- small quantity of >95% ethanol
- hand-held Tesla coil[1] or other high-voltage source
- safety glasses
- large tongs or ring stand (optional)

PROCEDURE

Place two to three milliliters of >95% ethanol (C_2H_5OH) in a 250-milliliter polyethylene bottle with two nails or screws through the opposite sides as shown in the illustration [1]. The nails or screws have pointed tips that come within a few millimeters of touching. Seal the bottle (not too tightly) with a cork or rubber stopper. Shake the bottle briefly to ensure that the liquid ethanol has come to equilibrium with its vapor. Then connect the nails to a high-voltage (a few kilovolts), low-current source. The resulting spark ignites the ethanol and blows the stopper off.

Alternately, bring a hand-held Tesla coil near one of the nails. Even with the second nail disconnected, there is enough capacitance between the nails to cause a spark to jump across the gap. You can increase this capacitance, causing a more intense spark, by connecting the second nail to ground or to any conductor with a large surface area such as a layer of aluminum foil on the outside of the bottle. For better visibility, hold the bottle with a pair of large tongs. An alternate arrangement uses a ring stand to support the bottle to keep it from tipping over with one screw through the bottom of the bottle in contact with the metal base of the ring stand.

---

[1] Available from Carolina Biological Supply Company, Educational Innovations, Fisher Science Education, and Sargent-Welch.

You can place a number of such bottles in series electrically, causing all the stoppers to blow off at the same time. Use larger bottles for more impressive explosions. Mention that automobiles can use ethanol as fuel and that the reaction in the cylinders of the engine is the same as the one illustrated here, with the spark provided by the spark plugs and the piston replacing the cork.

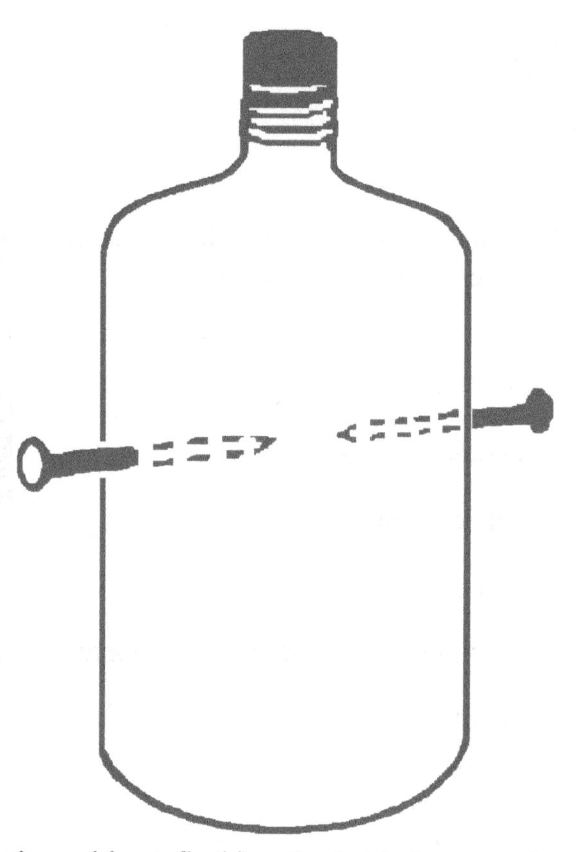

DISCUSSION

In this demonstration, some of the liquid ethanol evaporates and mixes with the air in the bottle. The reaction that takes place is $C_2H_5OH + 3O_2 \rightarrow 2CO_2 + 3H_2O$. Thus there is an increase both in the number of moles of gas (4 to 5) and in its temperature as a result of the energy released in the exothermic reaction. The increased pressure blows the stopper off the bottle. You cannot repeat the demonstration without flushing the bottle because the explosion consumes all the oxygen in the bottle. Blowing gently over the mouth of the bottle is usually sufficient to replenish the required air. To be sure, fill the bottle with water, and then empty it.

HAZARDS

The stopper is expelled with considerable force, and thus you should take care to avoid damage to light fixtures or other objects. Keep the area above the bottle clear of obstructions. A cork stopper is safer than a rubber one if it were to hit someone. Do not force the stopper on too tightly, lest the bottle explode. The voltage used to produce the spark should either be of a high frequency, as with the Tesla coil, or low current to avoid electrocution. Be careful not to tip over the bottle since the remaining liquid alcohol will spill and likely catch on fire. Wear safety glasses with this demonstration.

REFERENCE

1. B. Z. Shakhashiri, *Chemical Demonstrations*, The University of Wisconsin Press: Madison, WI, Vol 2 (1985).

2.24 Smoke Rings

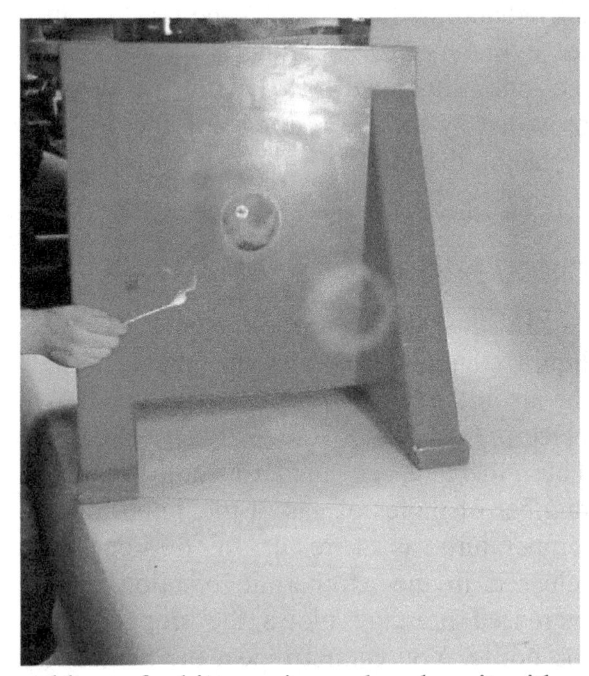

A cardboard box with a hole in one side produces smoke ring vortices.

MATERIALS

- cardboard or wooden box
- rubber or plastic sheet
- source of smoke
- spotlight or slide projector (optional)
- candle (optional)

PROCEDURE

Remove one of the six faces of a cardboard box of arbitrary size and replace it with a sheet of thin rubber or plastic. A plastic shower curtain, a trash bag, or even a sheet of cloth will suffice. Tape the sheet to the box to make it as nearly airtight as possible. Even more simply, use a large empty cereal box with the lid tightly sealed. Cut a circular hole about 10 cm in diameter in the cardboard side of the box opposite the sheet. Fill the box with smoke. A good way to do this is to paint the rim of the hole with an aqueous solution of titanium tetrachloride ($TiCl_4$) using a cotton swab. Alternately place two rags in the box, one soaked in household ammonia, and the other soaked in dilute hydrochloric acid, whose fumes combine to produce a harmless white smoke. You could instead fill the box with smoke from the combustion of some suitable material such as smoke paper, or, perhaps more safely, with chalk dust or talcum powder. An incense stick in the box also works well. If a spotlight or other bright light such as a slide projector is available, aim it at the hole in the box from across the room.

Place the box on its side, and tap the side with the sheet with your hand, as you would beat a drum. A smoke ring vortex should emerge from the hole and travel across the room. You can repeat the process as often as desired so long as some smoke remains. After a while the whole room becomes smoky, and visibility is impaired. The first few rings are usually the most dramatic. You can produce a fast-moving ring immediately after a slower one so that it catches up and overtakes the slower one by tapping rapidly twice, with the first tap lighter than the second. You can use two such vortex generators projected toward one another to study the collision and interaction of the vortices. You should point out that the vortices are there even in the absence of the smoke, whose purpose is to render them visible. You can illustrate this fact by blowing out a candle from across the room [1].

The construction of such a vortex generator makes an ideal home project for people of all ages because it is simple and safe as long as they do not burn the house down trying to make smoke. You can scale up the vortex generator to a very large size. You can encourage the audience to try making one and to investigate the effect of using holes of different sizes and shapes. Would a square hole make a square ring? What happens when the ring moves upward or downward?

DISCUSSION

It is easy to understand the mechanism whereby the vortex ring is generated [2–5]. When the air exits the hole, friction with the rim of the hole retards it, causing the air in the center to move forward faster than the air at the edge. If you imagine riding along with the vortex, the air at the center is moving forward and the air at the edge is moving backward. This leaves a region of reduced pressure in front of the ring at the edge and behind the ring in the center. The extra air at the front center circulates outward, and the extra air at the back edge circulates inward to equalize the pressure, forming the ring.

Vortex rings are quite common in nature. Flowing liquids form vortices when they flow too fast down a narrow channel, as can be seen by watching a fast-flowing river. Vortices off the wing tips of fast-moving aircraft are a hazard to other planes that inadvertently fly through them. Sometimes they are visible when a plane flies through humid air because the air drawn into the vortex core cools by expanding, and the humidity condenses to form a fog. They are caused by the high-pressure air beneath the wings circulating around the wing tips to the low-pressure region above the wings. Birds fly in a Vformation to take advantage of the lift from the vortex off the wingtip of the bird in front. Most people have seen a smoker make vortex rings by shaping the mouth in the form of a circle and exhaling gently. Scientists once thought (incorrectly) that vortex motion inside the atom controlled the interaction of atoms and their spectral emission.

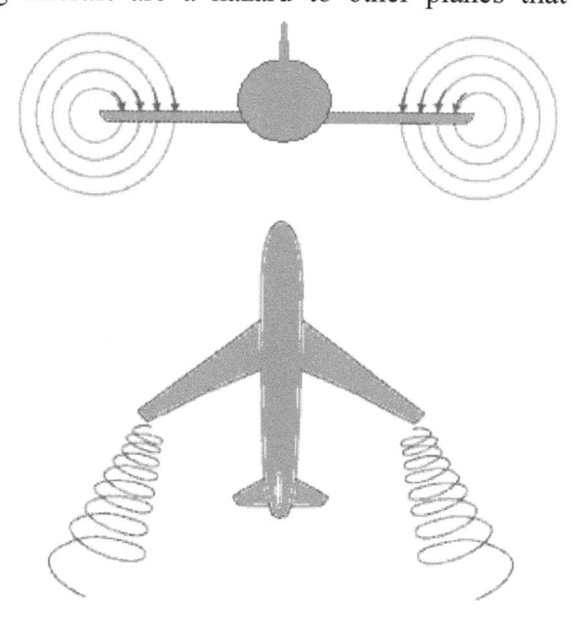

HAZARDS

The hazards are mostly in producing the smoke. Do not place ignited materials inside the box. When blowing out a candle from across the room, keep the candle well away from anything flammable.

REFERENCES

1. K. Bouffard, *Phys. Teach.* **38**, 18 (2000).

2. N. Didden, *Z. Angew. Math. Phys*. **30**, 101 (1979).

3. J. D. A. Walker, C. R. Smith, A. W. Cerra, and T. L. Doligalski, *J. Fluid Mech.* **181**, 99 (1987).

4. P. Orlandi and R. Verzicco, *J. Fluid Mech.* **256**, 615 (1993).

5. M. Nitsche and R. Krasny, *J. Fluid Mech.* **276**, 139 (1994).

2.25
Firehose Instability

A rubber hose connected to a source of compressed air dangles from a support and flails about in a chaotic manner.

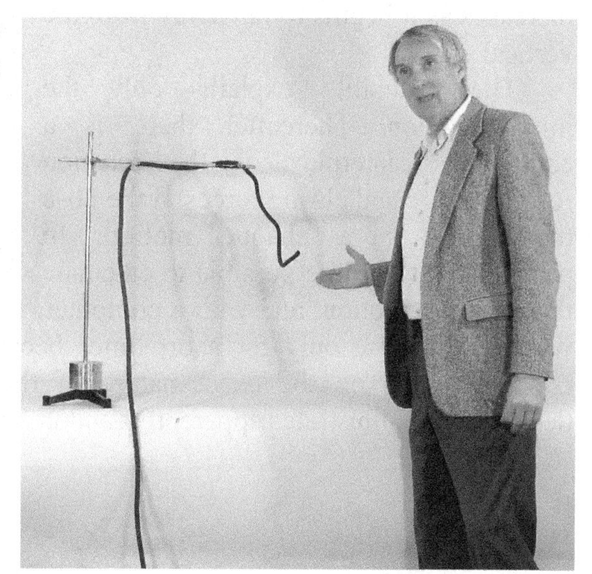

MATERIALS

- rubber hose
- source of compressed air
- ring stand or other support
- balloon (optional)
- pencil (optional)

PROCEDURE

Suspend a rubber hose from a support with one end open and about head level. Connect the other end to a source of compressed air. Ask the audience what would happen if they were to turn on a garden hose full blast without holding on to the end of it. Most people answer by flailing their arms wildly. Point out that this effect is called the "firehose instability," and that it can be dangerous with a large hose such as used by firefighters. Then turn on the compressed air to illustrate that the same thing happens with gases as with liquids. Point out that it is the same compressed air that one would use to inflate an automobile or bicycle tire. Try hoses of different diameters and with different nozzles to get the best effect for the length and pressure used. A similar phenomenon occurs when you release a balloon inflated with air. Show that at low flow rates, the hose moves with small, periodic oscillations, in contrast to the chaotic oscillations that occur at large flow rates. Explore the transition between the two types of behavior.

DISCUSSION

A hose with fluid coming out one end is unstable in the same way a pencil balanced on its point is unstable. If the hose were perfectly straight, the reaction force of the exhausted fluid would be along the axis of the hose and would simply compress it without any sideward motion. With a stiff hose or pipe, this is just what would happen. This condition is one of unstable equilibrium, just as a pencil balanced vertically on its point has all the force along its axis. However, if the hose bends slightly, there is a

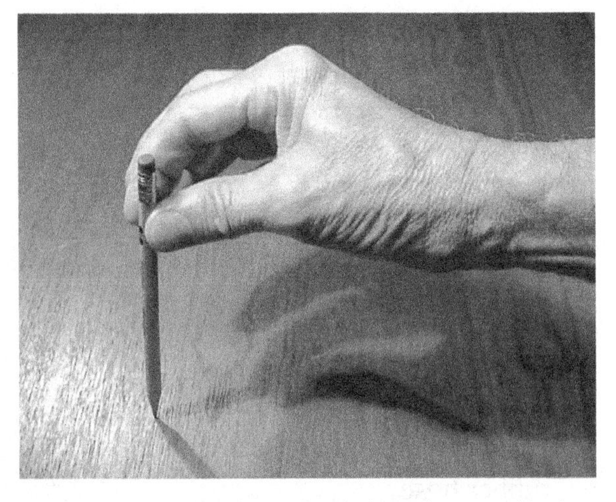

transverse component of the force that causes it to bend more, and the bend continues to increase. In the same way, if the pencil is not exactly vertical, there is a transverse component of the force that increases as it begins to fall away from the vertical.

The instability explains only the initial motion. Thereafter there is a complicated interplay of the reaction force, gravity and elastic forces in the hose that result in a chaotic motion. In principle, it would be possible to calculate the resulting motion, and with a computer, it might even be practical, but the prediction would be precise only for short times because of the sensitive dependence on initial conditions and on the exact nature of the forces. Like all chaotic processes, this demonstration illustrates apparently random behavior of deterministic systems [1].

HAZARDS

Hanging the hose above head level prevents it from hitting anyone in the face. Be sure to suspend it securely. Scaled-up versions of this demonstration using high-pressure gases and liquids and large hoses are impressive but entail additional obvious hazards.

REFERENCE

1. J. C. Sprott, *Chaos and Time-Series Analysis*, Oxford University Press: Oxford (2003).

2.26
Dripping Faucet

A dripping faucet illustrates periodic and chaotic behavior and the period-doubling route to chaos.

MATERIALS

- water faucet or valve with fine control
- water, preferably colored
- aluminum pie pan
- microphone, amplifier, and loudspeaker (optional)
- oscilloscope (optional)
- metronome (optional)
- Geiger counter and radioactive source (optional)
- magnifying glass or video camera (optional)

PROCEDURE

Place an aluminum pie pan about half a meter below a water faucet (“tap” in the UK) whose drip rate you can control. The pie pan makes the dripping water audible. Place a microphone against the pie pan and connect it to an amplifier and loudspeaker to make an impressive sound easily heard by a large audience. You can also connect the output of the microphone to an oscilloscope adjusted to a slow sweep rate (about 1 Hz) and set to trigger on the input sound. To improve visibility, use a video camera aimed at the point where the drops release. For added drama, play a recording of thunder before opening the faucet.

At a low drip rate (once per second, say) the dripping is periodic. Compare the dripping to the regular sound of a metronome. As the drip rate is slowly increased, there comes a point where a bifurcation occurs and the dripping suddenly changes to a period twice as long. The bifurcation is evident in the sound, which should change from “drip – drip – drip” to “drip drop – drip drop – drip drop.” Careful adjustment and patience is required to detect the bifurcation. At a slightly larger drip rate, a second bifurcation occurs (four times the period), and so forth, but these longer periods are difficult to observe. At the end of this infinite period-doubling sequence, the dripping is chaotic and has no detectable pattern. It should be easy to hear and observe on an oscilloscope the chaotic condition. The sound resembles the random clicks of a Geiger counter in the presence of a radioactive source, which you can demonstrate with the appropriate equipment. Point out that the clicks of the Geiger counter resemble the sound made from popcorn, in which the occurrence of a pop is unpredictable and the kernel of corn is irrevocably changed. Encourage the audience to explore the dripping faucet at home in their kitchen or bathroom.

DISCUSSION

The dripping faucet is one of the simplest and most common examples of the period-doubling route to chaos. It has been extensively studied [1–6], and an entire technical book (not a children's book) has been written about it [7]. Its simplicity makes it ideal for home experimentation.

The mechanism is as follows. There is a maximum size that a drop can have for its weight to be less than the surface tension that holds it together and keeps it attached to the rim of the faucet. When it reaches this critical size, the drop splits into two parts, one of which falls while the other remains behind. The drop left behind then begins to grow, and the process repeats. If the drop fills with water at a constant rate, the time required for it to reach the critical size is a constant, and the drops release periodically. However, the drop left behind will oscillate up and down a few times before coming to rest, and the inertia of this moving mass will either encourage or discourage the next release of a drop depending on the phase of the oscillation when the drop reaches its critical size. A drop released early will leave more water behind for the formation of the next drop and vice versa. It is reasonable that there might be a condition in which alternate drops are in and out of phase with the oscillation, respectively, and this is the condition that leads to period doubling. Chaos occurs when the drip rate is comparable to the oscillation frequency so that the phase of the oscillation is different each time a drop releases. The oscillation is easily visible if you observe the faucet closely, perhaps with a magnifying glass or video camera.

Water drops are a useful analogy to nuclear processes. There is a maximum size of the nucleus determined by a competition of the electrostatic repulsion between the protons and the short-range attractive nuclear force that behaves somewhat like surface tension. When a nucleus exceeds a certain critical size, it splits by the fission process with a large release of energy. The fission fragments are of varying size, determined in part by the phase of oscillations within the nucleus when the fission occurs.

HAZARDS

The only hazards are from spilled water and perhaps water damage to the microphone.

REFERENCES

1. P. Martien, S. C. Pope, P. L. Scott, and R. S. Shaw, *Phys. Lett. A* **110**, 399 (1985).

2. X. Wu, E. Tekle, and Z. A. Schelly, *Rev. Sci. Instr*. **60**, 3779 (1989).

3. X. Wu and Z. A. Schelly, *Physica D* **40**, 433 (1989).

4. H. Yepez, N. Nuniez, A. L. S. Brito, C. A. Vargas, and L. A. Vicente, *Eur. J. Phys.* **10**, 99 (1989).

5. R. F. Calalan, H. Leidecker, and G. D. Cahalan, *Comp. in Phys*. **4**, 368 (1990).

6. K. Dreyer and F. R. Hickey, *Am. J. Phys*. **59**, 619 (1991).

7. R. Shaw, *The Dripping Faucet as a Model Chaotic System*, Ariel Press: Santa Cruz, CA (1984).

3 Sound

Sound appears to be a topic distinct from motion and heat. However, we now understand sound to be an ordered motion of the molecules of the medium through which the sound propagates. The study of sound provides the opportunity to understand wave motion in preparation for the more abstract study of electromagnetic waves. The subject should be of particular interest to anyone who enjoys music [1–9].

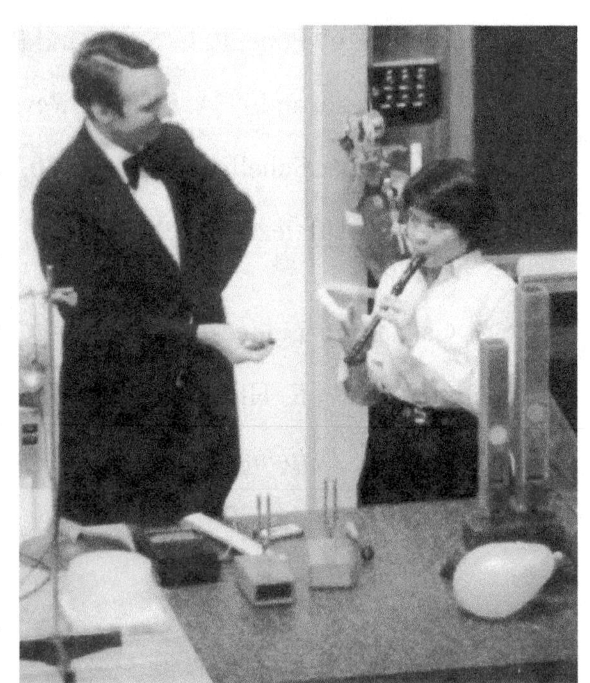

REFERENCES

1. H. E. White, *Physics and Music: The Science of Musical Sound*, Holt, Rinehart, & Winston: New York (1980).
2. A. Wood and J. M. Bowsher, *The Physics of Music*, John Wiley & Sons: New York (1981).
3. J. R. Pierce, *The Science of Musical Sound*, Scientific American Books, W. H. Freeman: New York (1992).
4. J. Askill, *Musical Sound: An Introduction to the Physics of Music*, Van Nostrand: New York (2001).
5. M. J. Moravcsik, *Musical Sound: An Introduction to the Physics of Music*, Paragon House Publishers: New York (2001).
6. J. G. Roederer, *The Physics and Psychophysics of Music: An Introduction* (3rd ed.), Springer-Verlag: New York (2001).
7. T. D. Rossing, F. R. Morre, and P. A. Wheeler, *The Science of Sound* (3rd ed.), Addison Wesley Publishing Company: New York (2001).
8. I. Johnston, *Measured Tones: The Interplay of Physics and Music* (2nd ed.), Institute of Physics Publishing: London (2002).
9. R. E. Berg and D. G. Stork, *The Physics of Sound* (3rd ed.), Pearson Prentice-Hall: Englewood Cliffs, NJ (2005).

Safety Considerations with Sound

Most experiments involving sound are relatively safe except for the hazards not directly associated with the sound as discussed elsewhere. However, it is possible for intense sounds to permanently damage the ear especially after prolonged exposure [1, 2], or if electronic amplification is used. Consequently, in the demonstrations described here, you should always keep the sound intensity at a level well below the threshold of pain (see table 3.1).

As a general rule, prolonged exposure to sound levels above 85 dB will cause slight hearing loss and above 90 dB will result in mild to moderate loss. A prolonged exposure to sound levels above 95 dB causes moderate to severe hearing loss. Above 100 dB, even short exposure can cause a permanent loss of hearing. Every 3 dB corresponds to a doubling of the intensity and hence to an approximate doubling of the perceived loudness.

The ear has an acoustic reflex that protects the inner ear from loud sounds in the same way the pupil of the eye protects the eye by contracting in the presence of bright lights. However, the reflex requires a few hundredths of a second to respond, and thus you cannot rely on it to protect the ear in the event of a short duration sound such as an explosion.

The damage is typically by way of tearing or ripping the microscopic hair cells of the cochlea. Such damage is usually only temporary unless the sound is frequent or sustained. Especially intense sounds are capable of rupturing the eardrum. The damage threshold is frequency-dependent, and the ear is most susceptible to damage by sounds of around 3,000 Hz, in part because the auditory canal is a closed tube having a resonance in this region. Prolonged exposure to sounds of a particular range of frequencies can permanently reduce the sensitivity of the ear to those frequencies.

The monitoring of sound levels requires an A-scale sound-level meter with a frequency response matched to the response of the ear or one capable of displaying the sound level for the entire audio frequency range [3]. You can provide protection by using earplugs and/or earmuffs. Cotton in the ears is not an effective protection unless the cotton is wax-impregnated.

REFERENCES

1. K. D. Kryter, W. D. Wood, J. D. Miller, and D. H. Eldredge, *Journ. Acoustical Soc. Am.* **39**, 451 (1966).

2. E. A. Lacy, *Handbook of Electronic Safety Procedures*, Prentice-Hall: Englewood Cliffs, NJ (1977).

3. A. P. G. Peterson and E. E. Gross, *Handbook of Noise Measurement*, General Radio: Concord, MA (1963).

Table 3.1

Sound intensities

Decibels (dB)	Intensity (W/m^2)	$\Delta p/p_0$	Example
0	10^{-12}	2×10^{-10}	Threshold of hearing
20	10^{-10}	2×10^{-9}	Whisper (at 1 m)
40	10^{-8}	2×10^{-8}	Mosquito buzzing
60	10^{-6}	2×10^{-7}	Normal conversation
80	10^{-4}	2×10^{-6}	Busy traffic
100	10^{-2}	2×10^{-5}	Subway
120	1.0	2×10^{-4}	Threshold of pain
140	100	2×10^{-3}	Jet engine

3.1
Wave Speed on a Rope

The difference in wave propagation speed for transverse waves on ropes of different masses and tensions is illustrated with a stick and two 3×5-inch index cards.

MATERIALS

- various ropes
- various weights
- two pulleys
- two 3×5-inch index cards (preferably of different colors)
- stick
- toy Slinky®[1] (optional)

PROCEDURE

Attach one end of a rope to the wall just above head level. Stretch the rope across the room, and give the free end a flick with your wrist. You can show that the speed of the wave on the rope increases with the tension in the rope but is independent of the shape of the wave. You can observe dispersion (change in the shape of the wave as it propagates).

Discuss the reflection of the wave at the fixed end. Note the inversion of the wave upon reflection from the fixed end. Explain that waves transport energy from one place to another even though the material through which the wave moves only oscillates back and forth in place. Point out that this demonstration illustrates how string musical instruments produce sound waves, but that sound waves are compressional waves rather than the

[1] Available from Carolina Biological Supply Company, Frey Scientific, PASCO Scientific, and Sargent-Welch.

transverse waves illustrated here. You can illustrate the difference using a toy Slinky® (the metal kind works best).

String two identical ropes across the front of the room. Make the tension different by placing one end of each rope over a pulley and hanging different weights on each with the other end of the ropes anchored securely in the opposite wall. Fold two 3×5-inch index cards (preferably different colors) in half and place one over each rope about a foot from the anchored end. Strike the opposite ends of the ropes simultaneously with a stick (such as a meter stick) while telling the audience to watch the cards. The card over the rope with the greatest tension will jump off the rope slightly, but noticeably, before the other.

To illustrate the variation of velocity with mass density, use two ropes with obviously different masses (say a quarter-inch-diameter and a half-inch-diameter rope). For ropes of the same material, the mass per unit length is proportional to the square of the diameter of the rope. Attach the ropes together and pass them around a single pulley so as to produce the same tension in each. Place cards over the ropes near the ends that attach separately to the wall, and pass the opposite ends around the pulley. Strike the ropes simultaneously with a stick as described above. The card over the smaller rope will jump off before the other.

You can combine the two demonstrations by using two ropes with different mass densities under different tensions, but with the same ratio of tension to mass density, in which case the cards should jump off simultaneously.

DISCUSSION

Fundamental to understanding the operation of musical string instruments is the velocity of a transverse, traveling wave on a string or rope under tension. Usually the velocity is so great that it is difficult to observe directly. The wave speed is equal to $\sqrt{T/\mu}$, where T is the tension and μ is the mass per unit length. Under the same tension, a string or rope of the same material but twice the diameter as another will propagate sound a factor of two slower. String instruments have thick strings (large μ) for the low notes and thin strings (small μ) for the high notes as well as some means to adjust the tension in the strings.

HAZARDS

String the ropes across the room above head level to prevent walking into them or tripping over them. Be careful to ensure that the weights do not fall on a foot if a rope breaks, comes loose from its anchor in the wall, or slips off its pulley.

3.2 Speed of Sound

The speed with which sound travels through the air is illustrated with a microphone and oscilloscope.

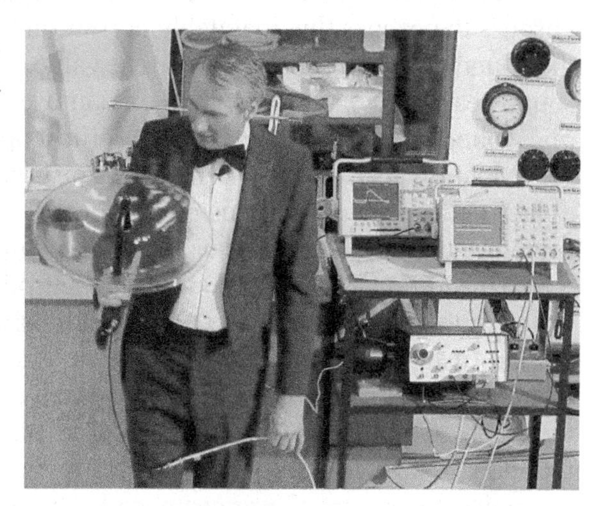

MATERIALS

- pulse generator and loudspeaker
- oscilloscope
- parabolic or other microphone
- flash lamp and kettle drum (optional)
- electric space heater (optional)
- two wood blocks, a meter stick and a stopwatch (optional)

PROCEDURE

Point out that when you talk, the sound of your voice appears to reach the ears of the listener instantaneously, but such is not the case. Ask if anyone knows the speed of sound. The correct answer is about 343 m/s (1,100 feet/second or 750 miles/hour) at room temperature. At the lower temperatures where jet airplanes fly, the speed is somewhat slower. At 30,000 feet, where the typical temperature is −44°C, the speed of sound is about 680 miles/hour. The speed of an object relative to the speed of sound in the medium through which the object is moving is called the Mach number (After Ernst Mach, 1838 – 1916) and was once thought to be a limit (the "sound barrier") that could not be exceeded by airplanes.

Offer to illustrate the speed of sound. Wrap your knuckles on the table rapidly to make a repetitive sound. Then produce the same sound with a pulse generator set to about 10 pulses per second and connected to a loudspeaker [1]. The pulses should be fairly narrow (about one millisecond). Adjust the oscilloscope to trigger on the pulses from the generator, and set the sweep rate for about 2 ms/division (20 ms full scale). If the oscilloscope will display two simultaneous waveforms, display the pulse waveform on one trace.

Then pick up a parabolic microphone connected to the other input of the oscilloscope. Ask the audience where they have seen such a thing. Many people have seen them along the sidelines at a football game. Ask why they might use such a thing. Some people may think it is to listen to what plays they are calling in the huddle. Point out that the more likely explanation is to pick up the grunts and groans and collision sounds to add drama to radio and television broadcasts.

Place the microphone close to the loudspeaker, and show that the sound is arriving nearly instantaneously. Then slowly back away. The pulse received by the microphone will get smaller and will move to the right on the oscilloscope, showing the time delay. Discuss the echoes that inevitably appear on the oscilloscope trace. It is easy to check that the sound speed has about the correct value. Explain that the pulse gets smaller when less sound reaches the microphone. You can use a volunteer from the audience to handle the microphone while you point to the oscilloscope.

Mention that this is the mechanism used by sonar to determine the distance to an object under water except using sound that reflects off the objects. Radar works by the same principle except that it uses electromagnetic (radio) waves rather than sound waves.

An additional discussion or way of introducing the idea is to ask how one can tell the distance to a lightning strike. After establishing that sound travels about 1,100 feet per second, ask how many feet are in a mile and then how many seconds it will take for sound to travel a mile. Flash a bright light, and count to five, at which point someone in the back of the room or offstage could begin pounding a kettledrum to simulate the thunder, or you could use a computer-recorded sound effect of thunder. You could use the noise to prompt the appearance of a surprise visitor. Point out that light does not travel instantaneously either, but that its speed is about a million times faster than sound.

With the microphone in a fixed position, use an electric space heater to heat the air between the loudspeaker and microphone. Show that the sound propagates faster, but not by much. The effect is small but noticeable. If you do not have a space heater, try using a hair dryer, although the noise from the blower may make it difficult to observe the sound from the loudspeaker.

For an outdoor project, find a building with a large wall about 100 meters away. Clap two wooden blocks together to make a loud sound, and record the time required to hear the echo using a stopwatch. Calculate the speed of sound from twice the distance to the wall divided by the time you record on the stopwatch. For better accuracy, clap the blocks in rhythm with the echo, with the echo occurring at the midpoint of successive claps. Count the number of claps in 20 seconds using the stopwatch, with the first clap at time zero. The speed of sound is then the distance to the wall times the number of claps divided by five.

This is one of a number of demonstrations for which it is good to point out the role of mathematics in physics and the importance of studying mathematics in order to understand the physical world. Ask the audience what the language of physics is, or more precisely, what language most accurately describes the laws of physics. You will probably get many amusing and revealing answers before someone thinks to suggest mathematics. Point out that mathematics is the language of nature itself and that everyone has learned to speak that language to some degree.

DISCUSSION

The speed of sound in air at 20°C is 343 m/s and is proportional to the square root of the absolute (in kelvins) temperature and independent of the pressure. The speed is also inversely proportional to the square root of the average molecular weight of the gas (see section 3.3), so that it travels about 2.7 times faster in helium, for example, whose molecular weight is 4, than in air, whose molecular weight is 29 (a weighted average of 28 for nitrogen and 32 for oxygen). Sound also travels in liquids and solids. The speed of sound in water is about 1,500 m/s, and it is about 5,000 m/s in a typical metal.

HAZARDS

There are no significant hazards with this demonstration provided you keep the sound level within reason. Do not trip over the cord that attaches to the microphone.

REFERENCE

1. R. S. Worland and D. D. Wilson, *Phys. Teach.* **37**, 53 (1999).

3.3
Breathing Helium and Sulfur Hexafluoride

The peculiar sound of one's voice after breathing helium or sulfur hexafluoride provides an amusing demonstration of the variation of the speed of sound with the density of a gas.

MATERIALS

- helium and sulfur hexafluoride[1]
- balloon
- funny hats (optional)
- organ pipe (optional)

PROCEDURE

You can dramatically illustrate the variation of the speed of sound with the density of a gas by breathing a low-density and a high-density gas and talking as you expel the gas from your lungs. Suitable gases are helium and sulfur hexafluoride. Xenon, though more expensive, can be used in place of sulfur hexafluoride. You can most conveniently breathe the gases by first filling a balloon from a gas cylinder with a pressure regulator set to about 5–10 psi. You should exhale the air from your lungs and then inhale the gas through the mouth as if sucking a straw. Then slowly exhale the gas while you talk. A duck hat for the helium and a cowboy hat for the sulfur hexafluoride add to the humor. Ask who has seen someone breathe helium or has personally done it.

You can also fill a balloon with sulfur hexafluoride and tie it off. It is distinctly heavier than an equivalent air balloon as can be illustrated by dropping them side-by-side. Since most people have heard the effect of helium on one's voice, it is best to start with that, and then breathe the sulfur hexafluoride by way of comparison and contrast. Mention that sulfur hexafluoride has the additional property that, since it is a high-density gas, it tends to remain in the lungs for a long time, and thus the effect on the voice persists much longer. Threaten the audience that you will sound funny for the rest of the lecture unless someone holds you upside down and empties out the gas. If there are no takers, bend over and exhale deeply a few times. This demonstration can also be done with a special guest breathing helium while you breathe the sulfur hexafluoride (or vice

---

[1] You can procure specialty gases and equipment from Matheson Tri-Gas, 166 Keystone Drive, Montgomeryville, PA 18936, 800-416-2505, http://mathesongas.com/.

versa) and carry on a dialog. You can describe one person as "being on a high" and the other "kind of low today." Try singing a duet for added amusement.

An especially instructive demonstration is to breathe a mixture of helium and sulfur hexafluoride in a proportion such that its density is close to the density of air. You can do this by first filling the balloon with helium and then adding just enough sulfur hexafluoride so that the balloon is slightly heavier than air (like an air-filled balloon). In principle, your voice should sound normal after breathing such a mixture, but in practice the helium rises first into the nasal cavity and is expelled, leaving behind the sulfur hexafluoride, which is exhausted at a slower rate, although the quantity of sulfur hexafluoride may not be sufficient to exhibit the effect. Such a demonstration shows that the density of the gas rather than its particular composition is the relevant variable.

If you are uncomfortable breathing these gases or if the gases you have available are not of high purity, you can do the demonstration instead by flowing the gases into organ pipes and observing the change in pitch.

DISCUSSION

The speed of sound in a gas is inversely proportional to the square root of its mass density, which in turn is proportional to its molecular weight at a given pressure and temperature (or number density). For helium, the speed is 2.7 times the speed of sound in air. For sulfur hexafluoride, the speed is 0.44 times the speed of sound in air. The speed of sound in air is 343 m/s or 750 mi/hr at 20°C and is proportional to the square root of the absolute (kelvin) temperature.

The voice sounds different because the natural resonant frequencies of cavities in the head increase in proportion to the speed of sound [1]. Of course the resonant frequencies of these cavities are only one ingredient in the way your voice sounds. The gas does not alter the frequency of the vocal cords (about 140 Hz for males and 230 Hz for females on average when speaking, plus higher harmonics) nor does it affect how fast you talk. The vocal tract acts as a filter for the sounds produced by the vocal cords in a manner similar to the treble and bass controls on a stereo. However, there is a feedback mechanism from the ears through the brain that may alter the way some people manipulate their vocal cords. In fact, this demonstration works much better for some people than for others.

Most people have heard the distinctive yet highly amusing sound that results when one breathes helium. Deep-sea divers often breathe an atmosphere that is largely helium so that they can exist in a high-pressure environment without exceeding the body's tolerance for nitrogen and oxygen. Most people have not heard the even more amusing sound that results from breathing sulfur hexafluoride.

HAZARDS

The major danger in performing this demonstration is hypoxia from not getting enough oxygen while you breathe the gas. Thus you should not immediately repeat the demonstration but rather wait a minute or so before breathing more gas. It is a good idea to take a few deep breaths before and after each demonstration. Even safer would be to breathe heliox, a diving-gas mixture of helium and oxygen designed to prevent the bends. You should point out to the audience that breathing most gases (even some air) is potentially harmful. Helium and sulfur hexafluoride are two of a small number of gases with which you can safely do this demonstration. The substitution of other gases, especially flammable gases such as hydrogen, is not recommended. The gas should be of high purity (research grade) to avoid contaminants such as pump oil or other toxic gases such as compounds of sulfur and fluorine that may be harmful. Rinse the inside of the balloon with water to remove the fine powder (usually cornstarch) that is often present to prevent the rubber (latex) from sticking to itself. This can be done most effectively by first turning the balloon inside out. If the gas has an odor, do not inhale it. Do not do this demonstration or encourage others to do it using the helium from balloons bought at a fair or circus because that gas is insufficiently pure and the inside of the balloons may not be sufficiently clean.

REFERENCE

1. A. H. Cromer, *Physics for the Life Sciences*, McGraw Hill: New York (1977).

3.4 Bell in Vacuum

An electric bell in a jar makes a sound that decreases in intensity as the air is evacuated from the jar.

MATERIALS

- electric doorbell or hand-held bell
- battery or power supply (for the electric bell)
- bell jar[1]
- vacuum pump
- pressure gauge (optional)
- sound meter (optional)
- metronome (optional)

PROCEDURE

You can demonstrate the necessity of a medium for the propagation of sound by using an electric bell (a doorbell type) or a metronome in an evacuated bell jar [1]. (The bell jar is so named not because it contains a bell but because it is usually shaped like a bell with a rounded end and straight sides.) You can bring the electrical leads for the bell out of the bell jar through vacuum feedthroughs. Transmission of sound by the electrical leads requires that they have a small cross section and are rather long and coiled up. You can lead them through a rubber stopper in a hole in the glass to dampen the transmitted sound even more. Suspend the bell by its leads so it does not touch the bell jar. Turn the bell on and evacuate the bell jar to a pressure in the millitorr range with a mechanical vacuum pump while the sound slowly diminishes until it is almost inaudible. With a sound meter and pressure gauge, you could record the sound intensity as a function of pressure.

In a simpler version, mount a small bell with a clapper inside a jar small enough to hold in your hand and shake. Equip the jar with a rubber stopper, hose, and shut-off valve. In this variation, evacuate the jar before the lecture, and then shake the jar while

---

[1] Available from American 3B Scientific, Carolina Biological Supply Company, Fisher Science Education, Frey Scientific, Nasco, Sargent-Welch, and Science Kits & Boreal Laboratories.

air is slowly admitted, causing the sound to become audible. The sound is not very loud, but by holding the jar close to a microphone, the audience can easily hear it.

In a more elaborate version, mount a microphone inside the bell jar and connect its amplified signal to a loudspeaker and oscilloscope [2]. Note, however, that some microphones do not work properly in a vacuum and may even be destroyed. In addition, the microphone is likely to pick up electrical noise from the bell, and so a metronome in the bell jar is better for this variant.

DISCUSSION

This experiment was first performed by the science philosopher Athanasius Kircher (1602 – 1680) and later by the Irish physicist Robert Boyle (1627 – 1691) [3]. The sound propagation should cease when the mean free path for air molecules becomes longer than a few centimeters. At room temperature and 760 Torr of pressure (1 atmosphere), the mean free path is about 2×10^{-5} cm and is inversely proportional to the pressure. Thus you should ideally reduce the pressure to about 10 millitorr to reduce the sound to the lowest possible level. Such pressures are hard to achieve with typical mechanical vacuum pumps.

This demonstration is actually a bit of a swindle since the dominant effect is the impedance mismatch between the bell and the air in the jar, and between the air and glass, that causes reflection rather than transmission of the sound at the interfaces. The impedance of the air is proportional to the square root of the product of its pressure and density. This effect causes the demonstration to work even with an only partially evacuated jar.

HAZARDS

The only significant hazard with this demonstration is implosion of the bell jar. A jar designed for vacuum use (heavy flint glass), should be safe if you avoid striking it with a heavy object or otherwise cracking it. The bell should be of the type that operates from a low voltage (0–6 volts) so as not to pose an electrical hazard. A variation of the demonstration, in which you place a battery inside the bell jar to operate the bell, is not recommended since the battery may explode when exposed to a vacuum. Make sure the belt of the vacuum pump has a guard, or place the pump out of reach.

REFERENCES

1. R. B. Lindsay, *Am. J. Phys.* **16**, 371 (1948).

2. D. Han, *Phys. Teach.* **41**, 278 (2003).

3. R. Boyle, *New Experiments Physico-Mechicall, Touching the Spring of the Air and Its Effects*, Oxford University Press: Oxford (1660).

3.5
Doppler Effect

A reed mounted on the end of a rotating arm produces a tone whose pitch wobbles up and down as the arm rotates.

MATERIALS

- sound reed
- motor with rotating arm[1]
- strobe light (optional)
- smoke detector or tuning fork and rope (optional)
- loudspeaker with long leads and audio oscillator (optional)
- tape recording of train whistle or car horn (optional)

PROCEDURE

You can illustrate the Doppler effect (named after the Austrian physicist Christian Johann Doppler, 1803 – 1853) with a device consisting of a reed mounted on the end of an arm spun by a motor with suitable gears to control the speed [1]. The rotation causes air to pass over the reed, producing a sound whose pitch wobbles up and down as the reed successively moves toward and then away from the observer. It helps if you can adjust the speed of rotation. If you use a microphone for lecturing, turn it off or place it on the side of the apparatus toward the audience to reduce the interference between the direct and amplified sounds. You can view the spinning arm with a strobe light to freeze its motion. Alternate techniques for exhibiting the Doppler effect are also available [2–5]. You can swing a smoke detector wired so as to emit a continuous tone around on the end of a rope. Even more simply, swing a tuning fork around on the rope after striking it with a mallet, or with the heal of your shoe if you do not have a mallet. A loudspeaker connected to an audio oscillator with long leads will display a noticeable Doppler effect when moved rapidly toward or away from the audience. For example, a source tuned to a frequency of about 330 Hz should shift by about 5 Hz at a speed of 5 m/s (about 11 miles per hour), which you should be able to achieve briefly.

You can simulate the effect verbally, but a tape recording of a train whistle or a car horn is more effective. You can find recordings of the Doppler effect on sound-effect

[1] Available from Fisher Science Education, Sargent-Welch, and Science Kits & Boreal Laboratories.

albums at record libraries or on the web, or make one yourself with a tape recorder and automobile on a country road. This makes a good weekend family project for students who can then share their results with their classmates [6].

You can quantify this demonstration using a microphone and spectrum analyzer [7]. Software is available on the web for making sonograms using the sound card provided with most computers. You can use the measured Doppler effect from a sound source moving at a known speed to calculate the speed of sound [8].

DISCUSSION

The frequency heard by a fixed observer for a source in motion is

$$f = f_0 v/(v \pm v_s)$$

where f_0 is the frequency of the source, v_s is its velocity, and v is the velocity of sound (about 343 m/s in air at 20°C). The plus sign holds for motion away from the observer, and the minus sign holds for motion toward the observer. Thus if you want to produce a ±5% frequency shift for a reed at the end of a 25-cm-long arm, a rotation speed of about 650 rpm is required. The observed frequency versus time for an object emitting a sound of frequency f_0 on the end of an arm of radius R rotating with a period T is [7]

$$f = \frac{f_0}{1 + \frac{2\pi R}{vT}\cos\left(\frac{2\pi t}{T}\right)}$$

The Doppler effect was tested by the Dutch meteorologist Christophorus Heinrich Diedrich Buys Ballot (1817 – 1890) in a bizarre experiment in Holland a few years after Doppler derived his formula in 1842 [9]. For two days a locomotive pulled a flat car back and forth at different speeds. On the flat car were trumpeters sounding a particular note. On the ground, musicians with a sense of absolute pitch recorded the note as the train approached and receded.

There are many examples in nature of the Doppler effect. In addition to the usual train whistle and automobile horn, the Doppler effect also occurs for electromagnetic waves, but since the speed of light is much greater than the speed of sound, the effect is much smaller for a given source velocity. Even so, the police use Doppler radar to catch speeders. Doppler radar is used to measure the speed of baseballs and tennis balls. Air-traffic radars use the Doppler effect to discriminate against stationary targets and to detect wind shear. The Doppler effect is also important in the reception of radio signals from Earth-orbiting satellites. The Doppler effect is used to determine the distance of galaxies [10, 11] (the Hubble constant) and in the Mössbauer effect [12–14]. The Doppler effect of starlight is one method for deducing the presence of planets orbiting distant stars through the back and forth motion that their orbit causes on the motion of the star [15, 16].

HAZARDS

The main hazard in this demonstration is coming into contact with the rapidly rotating arm or the reed flying off the arm if it is not securely attached. Use a switch and speed control to permit operation from a safe distance. The apparatus should not contain parts that could come loose and fly out into the audience. If you measure the Doppler effect with a moving car, it is best to have three people in the car, one to drive, one to read out the speed, and a third to sound the horn.

REFERENCES

1. H. A. Robinson, Ed., *Lecture Demonstrations in Physics*, American Institute of Physics: New York (1963).
2. F. S. Crawford, *Am. J. Phys.* **41**, 727 (1973).
3. H. F. Meiners, *Physics Demonstration Experiments*, Vol I, The Ronald Press Company: New York (1970).
4. J. S. Miller, *Physics Fun and Demonstrations*, Central Scientific Company: Franklin Park, IL (1974).
5. V. Mallete, *Phys. Teach.* **34**, 126 (1972).
6. M. M. F. Saba and R. A. S. Rosa, *Phys. Teach.* **39**, 431 (2001).
7. M. M. F. Saba and R. A. S. Rosa, *Phys. Teach.* **41**, 89 (2003).
8. R. Gagne, *Phys. Teach.* **34**, 126 (1996).
9. D. Filkin and S. Hawking, *Stephen Hawking's Universe: The Cosmos Explained*, Basic Books: New York (1997).
10. E. Hubble, *Proc. Natl. Acad. Sci.* **15**, 168 (1929).
11. E. Hubble and M. Humason, *Astrophys. J.* **74**, 43 (1931).
12. A. Abragam, *L'effet Mössbauer et ses Applications*, Gordon and Breach: Paris (1964).
13. G. M. Bancroft, *Mössbauer Spectroscopy: An Introduction for Inorganic Chemists and Geochemists*, John Wiley & Sons: New York (1973).
14. R. V. Pound and G. A. Rebka, *Phys. Rev. Lett.* **3**, 439 (1959).
15. G. W. Marcy, *Nature* **39**, 127 (1998).
16. M. C. LoPresto and R. McKay, *Phys. Teach.* **42**, 208 (2004).

3.6 Flame Pipe

A pipe several meters long filled with natural gas and connected to a loudspeaker produces a flame whose height varies with position along the length of the pipe.

MATERIALS

- specially constructed pipe
- source of natural gas, methane, or propane
- matches
- loudspeaker
- audio amplifier (a few watts)
- sine-wave generator
- meter stick (optional)
- pulse generator, microphone, and oscilloscope (optional)

PROCEDURE

A brass or copper pipe about 5 cm in diameter and 2 meters long contains several hundred small-diameter holes in a line along the top of the pipe (also called a "Rubens tube"). A downspout pipe will also work. Seal one end of the pipe and fit it with a nozzle through which you can admit natural gas, methane, or propane. On the other end of the pipe, mount a loudspeaker, making a reasonably tight seal. Connect the loudspeaker to the output of an audio amplifier whose input you connect to a sine-wave generator of variable frequency.

Turn on the gas and wait a few seconds for the air to be expelled from the pipe. Then ignite the gas with a match where it exits the holes along the pipe. This produces a nearly continuous wall of flame several meters long. A considerable throughput of gas is desired, and the effect is best when viewed in subdued illumination. Adjust the frequency and amplitude of the sound emitted by the loudspeaker until you see a clear standing wave pattern in the flame. At certain frequencies, the wave will resonate in the pipe, and the required amplitude will be quite small. More amplitude is usually required as the frequency is increased, but you should be careful not to burn out the loudspeaker. As you raise the frequency, the distance between adjacent wave crests will decrease. You can also do the demonstration with other wave shapes such as square waves or the waves corresponding to speech and music.

You can use the same apparatus, without the flame, to demonstrate the phase change when a sound wave is reflected [1]. Connect the loudspeaker to a pulse generator that produces periodic pulses about 200 μs in duration. Connect a microphone to an oscilloscope triggered by the pulse from the pulse generator. Move the microphone along the length of the pipe just above the row of holes. You should see the pulse propagating along the pipe and then reflected back, with the reflected pulse having the same polarity as the incident pulse when the far end of the pipe is closed and the opposite polarity when the far end is open. Note that the reflected pulse has nearly the same amplitude as the incident pulse. This variant also allows you to measure directly the speed of sound in the pipe with various gases.

DISCUSSION

The fundamental principle illustrated in this demonstration [2–12] is the compressional wave nature of sound. Where the average pressure is greatest, the gas emitted through the holes is greatest, and the flame is most intense. The flame intensity is low where the wave displacement is largest as a result of the Bernoulli effect (see section 2.2). The information resembles that displayed on an oscilloscope but is more direct and less mysterious.

The variation of wavelength with frequency (or pitch) of the sound is easily illustrated. With a meter stick you can show that the product of frequency and wavelength is a constant equal to the speed of sound (about 460 m/s in methane at 20°C). Note that the distance between the nodes of the flame is half a wavelength.

This demonstration also illustrates the resonant frequencies of a closed organ pipe. The closed end of the pipe is a displacement node (the air cannot move there) and thus is a pressure antinode (maximum) since the displacement and the pressure are 90° out of phase with one another. Conversely, the end driven by the loudspeaker is a displacement antinode and a pressure node.

A resonance will occur in the pipe when the length of the pipe is an odd multiple of a quarter wavelength. At resonance, the pressure variation within the pipe becomes quite large, and a relatively low sound intensity results in a large variation in flame intensity. As the pipe heats up, the sound speed changes (proportional to the square root of the temperature in kelvins) and a slight readjustment of the frequency is required to maintain resonance. Variation in the air-to-gas ratio also causes the resonance to drift.

HAZARDS

Place the flame well away from any materials that it might ignite. The pipe becomes quite hot after prolonged operation and requires some time to cool off. If you light the flame too quickly after you turn on the gas while there is still air in the pipe, the flame could be sucked into the pipe, destroying the loudspeaker. Be careful not to inadvertently extinguish the flame and fill the room with gas. The substitution of other flammable gases such as ethane or hydrogen is not recommended.

REFERENCES

1. D. Potter, *Phys. Teach.* **41**, 12 (2003).
2. R. J. Stephenson and G. K. Schoepfle, *Am. J. Phys.* **14**, 294 (1946).
3. J. L. Underfer, *Phys. Teach.* **4**, 81 (1966).
4. R. Coleman, *Phys. Teach.* **13**, 556 (1975).
5. M. Iona, *Phys. Teach.* **14**, 325 (1976).
6. R. P. Bauman and D. Moor, *Phys. Teach.* **14**, 448 (1976).
7. T. D. Rossing, *Phys. Teach.* **15**, 260 (1977).
8. S. Trester, *Phys. Teach.* **15**, 426 (1977).
9. G. W. Ficken and F. C. Stephonson, *Phys. Teach.* **17**, 306 (1979).
10. G. F. Spagna, *Am. J. Phys.* **51**, 848 (1983).
11. G. Ficken and F. Stephenson, *Am. J. Phys.* **54**, 297 (1986).
12. G. F. Spagna, *Am. J. Phys.* **54**, 1146 (1986).

3.7 Oscilloscope Waveforms

An oscilloscope displays the waveforms of various musical instruments showing the effect of frequency and wave shape on the sound.

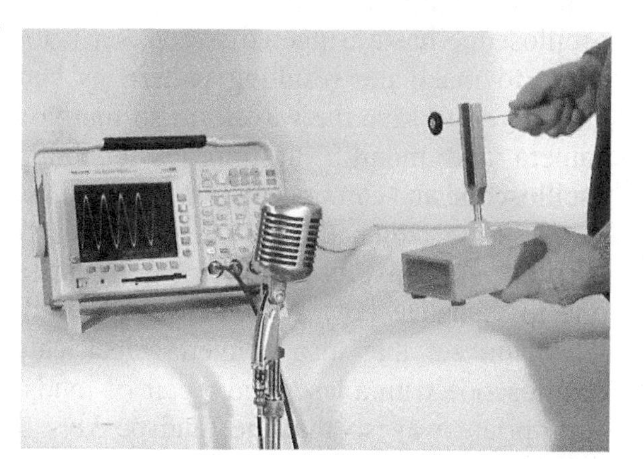

MATERIALS

- oscilloscope
- microphone
- various musical instruments and other sound sources
- tuning forks[1] (optional)
- strobe light (optional)
- beaker of water (optional)
- electronic keyboard (optional)
- function generator (optional)
- noise generator (optional)
- amplifier and loudspeaker (optional)
- inexpensive transistor radio (optional)
- corrugaphone[2] (optional)
- 1-cm-diameter aluminum rod at least a meter long (optional)
- toy Slinky® (optional)
- soap bubble solution (optional)

---

[1] Available from American 3B Scientific, Carolina Biological Supply Company, Frey Scientific, PASCO Scientific, Sargent-Welch, and Science Kits & Boreal Laboratories.
[2] Available from Arbor Scientific, Educational Innovations, and PASCO Scientific.

PROCEDURE

Connect an oscilloscope with a large screen directly to a microphone or to the output of an audio amplifier or function generator to illustrate the waveforms produced by various sounds. Set the sweep rate to about 200 Hz (5 milliseconds per sweep). This rate is fast enough to prevent flicker but slow enough to observe the low frequencies. If the oscilloscope has a triggered sweep, set it to "internal" and set the trigger level to a low value to make the resulting pattern as stationary as possible. A storage oscilloscope allows you to save the waveform so that you can discuss it. In a large room, use a video camera and monitor to improve visibility of the oscilloscope, in which case the oscilloscope and camera can be backstage and connected to the PA and video monitor.

Ask the audience if they recognize this device and what it is called. Explain that the oscilloscope converts the electrical signal from a microphone into a picture that allows one to visualize how a sound wave would look if you could see it. It is a bit like a television set that also converts electrical signals into a picture. You can make an oscilloscope with a laser and a pair of loudspeakers with mirrors mounted on them in an appropriate way so that the loudspeakers deflect the light beam in two perpendicular directions, allowing you to display the pattern on the wall for easy viewing. Show the complicated waveform corresponding to human speech.

You can produce sine waves with tuning forks or organ pipes or with an electronic function generator. Point out that sine waves are the purest type of wave, consisting of only a single frequency. You could first show the sounds corresponding to sine waves of different frequencies. Explain how a tuning fork works. Strike one tuning fork, and let the audience hear the sound. Explain that you are converting the mechanical energy of the mallet into an ordered mechanical motion of the tuning fork, which then produces sound energy that eventually dissipates as heat. With a sufficiently large tuning fork, you can actually see the vibrations of the arms (when the frequency is about 10 Hz or lower), or you can view the vibrations with a strobe light. Alternately, lower the tuning fork into a beaker of water. It will quickly stop vibrating, but will make a gentle spray of water while doing so.

Then hold up a second larger tuning fork and ask whether it will produce a higher or lower pitch before striking it. Point out that the lower pitch has a longer period as shown on the oscilloscope. The wavelength of audible sound ranges from about a centimeter up to a few meters. Anyone in the audience with perfect pitch should be able to identify the musical notes (see table 3.2). The way in which musical scales are constructed makes an interesting digression [1, 2]. The standard of A-440 is not universally accepted, nor is the equal-tempered scale. Some scientific manufacturers once adopted a standard of 256 (2^8) for middle C, but musicians ignored it.

With a function generator, you can show the frequency limits of hearing (approximately 20 Hz to 20 kHz). An effective way to do this is to start at a frequency that everyone can hear such as 5 kHz. Ask the audience to raise their hands as long as they are able to hear the sound. Then slowly increase the frequency. When the frequency rises to about 20 kHz, depending on the intensity, only the children's hands will remain raised. The high-frequency response of hearing deteriorates with age, a medical condition called "presbycusis."

Musical instruments such as a violin and a trumpet have distinctive waveforms [3–5]. You could show examples of string, wind, and percussion instruments. You can illustrate frequency modulation (vibrato) and amplitude modulation (tremolo). If someone in the audience can play the instruments, it adds a touch of interest to have them do so. Alternately, an accomplished musician can be planted in the audience who when called upon pretends minimal skills until suddenly launching into "The Flight of the Bumblebee" or some similarly intricate piece. Another idea is to have a musical quartet come onstage and explain the operation of each instrument before having them play a short piece.

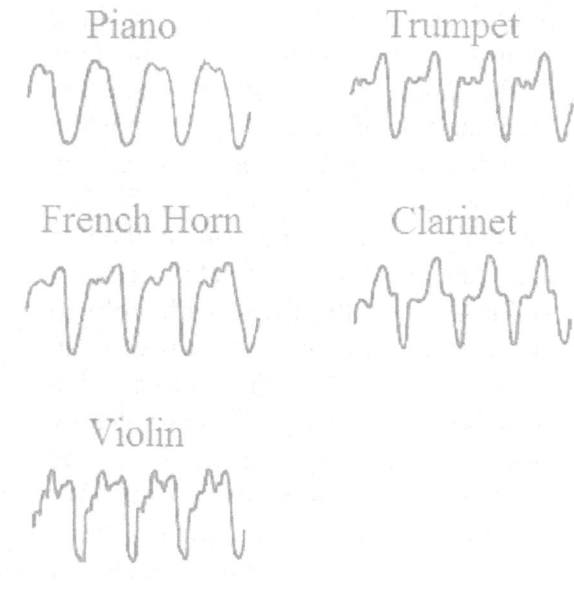

While you have a wind instrument handy, you can demonstrate that sound is a wave that does not result in a net motion of the air by dipping the end of the instrument into soap bubble solution. You can blow the horn without popping the bubble [6].

Modern electronic keyboards can emulate a wide variety of musical instruments. Demonstrating the difference in waveform between a real instrument and a synthesized instrument can be quite interesting. Playing a chord on the keyboard shows how the waveform becomes more complex as more frequencies are added. The complex waveform required for speech and music is especially dramatic. Roughly half of the intelligibility of speech is contained in the frequency range of 1,000 – 4,000 Hz, although only about 4% of the energy is in that range. The human voice has a distinctive waveform that varies from person to person, even when singing the same musical note or voicing the same sound. You might try illustrating this difference using two or more volunteers from the audience.

You can produce square waves and other waveforms with a function generator connected to an amplifier and loudspeaker as well as to an oscilloscope. You could illustrate the different sounds produced by sine waves and square waves of the same frequency. The difference results from what musicians call "overtones" and physicists call "harmonics." Harmonics are integral multiples of the fundamental frequency, but overtones may or may not relate to the fundamental in a simple way. In most string and wind instruments, the overtones form a harmonic series, but in percussion instruments such as the drum, the overtones are more complicated and less "harmonious" (not a single discernible note). The combination of overtones is what gives each musical instrument its characteristic quality, or timbre.

This demonstration offers the opportunity to exhibit unusual musical instruments and other sources of sound. You might try blowing over the top of a soft drink bottle and subsequently drinking from it to change its pitch. You can show that the pitch of the sound made by tapping the side of the bottle goes up while the pitch made by blowing over the top goes down as you drink from it [7]. The same thing can be done with a flexible plastic tube (1/2-inch diameter Eastman Poly-Flo®) partially filled with colored water, fruit juice, or wine, which you can drink when done. Cut the bottom off a Coke® bottle, and play different notes by blowing over the top while it is dipped different depths into a beaker of water. Wine glasses struck with a small mallet or rubbed along the rim with a clean, grease-free finger emit good sine waves. Music has been written for such a glass harmonica. Other possibilities include a bent saw blade, whistles, horns, and sirens.

A long, hollow pipe, open on both ends, lowered over a Bunsen burner (sometimes called a "hoot tube") will emit a sound [8], as was first shown by Sir Charles Wheatstone (1802 – 1875) in a series of public lectures at the Royal Institution in London [9]. If the pipe contains a heated screen,[3] the sound can persist as long as 30 seconds after you remove it from the flame. The sound stops if you hold the pipe horizontally, allowing you to "pour the sound out of the pipe."

You can use a corrugated plastic tube, called a "corrugaphone," "Bloogle Resonator," or "Hummer," to produce a variety of whistling sounds when you spin it around over your head [10]. The frequencies are harmonics of the fundamental organ-pipe mode that are individually preferentially excited depending on the speed of rotation. It is hard to excite the fundamental and even the second harmonic, but the higher harmonics are easily excited [11].

An especially impressive demonstration uses an aluminum rod about a centimeter in diameter and at least a meter long [12]. Hold the rod at the center in one hand. You can find the center quickly by balancing the rod on your finger or by placing a piece of masking tape there beforehand. Put powdered resin on the fingers of your other hand and stroke the rod along its length to excite longitudinal vibrations.

[3] Available from Educational Innovations.

These vibrations have a high pitch and vibrate for a long time after you stop stroking it, and are nearly sinusoidal. You can stop the sound abruptly by touching one of the antinodes at either end of the rod. Then grip the rod at a point one-quarter of the length from one end and repeat the stroking to excite the second harmonic. It helps to mark the appropriate spots with masking tape. More simply, hold the rod vertically and tap it on a hard floor to excite the vibrations or strike the upper end with a hammer if you do not have a hard floor. You can quantify your observations by measuring the frequency of the oscillation and using the fact that the speed of a longitudinal sound wave in aluminum is about 5,000 m/s.

You can compare these longitudinal vibrations with the transverse vibrations that occur when you strike the rod with a mallet from the side. Use a toy Slinky® to illustrate transverse vibrations (the metal kind works best). The transverse vibrations are at a much lower frequency unless the rod is very short and stiff as in a xylophone.

You can illustrate that the perceived loudness of a sound depends on its wave shape. The square wave sounds louder than a sine wave of the same frequency because of the harmonics, even when you lower the amplitude of the square wave by $\sqrt{2/\pi}$ to account for its larger average power. Furthermore, the perceived pitch of a sound depends slightly on its intensity [13]. Usually the pitch change is downward at low frequencies and upward at high frequencies as the loudness increases, but the effect varies considerably with the individual [14]. Furthermore, the effect is noticeable only for sine waves [15, 16]. You can demonstrate these perceptions with a pair of function generators and a switch to alternate between the two. If you sound the sources simultaneously, the presence of beats can obscure the effect.

You can connect an inexpensive transistor radio to the oscilloscope to illustrate the distortion that results when you turn up the volume too high and flatten the waves due to saturation of the transistor amplifiers. This could lead to a discussion of high-fidelity audio equipment and the need for high-power amplifiers to minimize distortion due to the generation of harmonics. If the radio has a tone control, you can show the effect of increasing and decreasing the bass and treble.

By tuning an FM radio or television between stations, you can demonstrate what noise sounds like. This noise, and the accompanying video image, is in part a remnant of the three-degree background radiation from the Big Bang that occurred about 13.7 billion[4] years ago. There are many types of noise, such as white noise (all frequencies present in equal amounts), $1/f$ noise (sometimes called "pink noise"), $1/f^2$ noise (sometimes called "brown noise" since it is characteristic of Brownian motion), and others. Commercial noise generators are available that produce the various types. Alternately, you can simulate various kinds of noise by filtering white noise through an audio equalizer (not adjusted to equalize). You can produce brown noise from white noise by integration using an RC low-pass filter.

DISCUSSION

It is interesting to note that the Fourier frequency components determine the shape of a waveform. The phase as well as the amplitude of the individual frequency components will affect the shape of the waveform, but since the ear is insensitive to the relative phases of the components of a wave [17], waveforms of quite different shapes may sound identical to the ear. Realize that displaying sound waves on an oscilloscope gives the false impression that sound is a transverse wave, like waves on the water, rather than a longitudinal compressional wave.

The ear is a remarkably sensitive detector of sound [18]. The threshold of hearing is about 10^{-12} W/m^2, which corresponds to a pressure variation of less than one part in a billion and a displacement of the air molecules by an amount less than their size. The sensitivity is almost enough to detect the fluctuations in air pressure due to the random thermal motion of the air molecules. The ear accomplishes this feat through its construction, which amplifies the pressure by a factor of about 100, corresponding to an increase in sound intensity by a factor of 10^4. At the other extreme, the ear can tolerate sounds as intense as 1 W/m^2. The ear accommodates the wide range of intensities by exhibiting an approximately logarithmic response. It is for that reason that sound intensities S are usually expressed in units of decibels (dB)

$$S = 10 \log_{10}(I/10^{-12})$$

where I is the intensity in W/m^2 and the factor of 10^{-12} represents approximately the threshold of hearing. Thus a sound of 0 dB is at the threshold of hearing, and a painfully (and potentially dangerous) loud sound is 120 dB.

HAZARDS

There are no precautions other than keeping the sound intensity at a safe level.

---

[4] This is the U.S. billion of 1,000 million or 10^9.

REFERENCES

1. J. Backus, *The Acoustical Foundations of Music*, W. W. Norton & Company: New York (1969).

2. E. E. Helm, *Scientific American* **217**, 93 (Dec 1967).

3. H. Lineback, *Scientific American* **184**, 52 (May 1951).

4. C. A. Culver, *Musical Acoustics*, McGraw-Hill: New York (1956).

5. N. H. Fletcher and T. D. Rossing, *The Physics of Musical Instruments*, Springer-Verlag: New York (1991).

6. W. G. Hults, *Phys. Teach.* **18**, 671 (1980).

7. E. Van den Berg and R. Van den Berg, *Phys. Teach.* **36**, 356 (1998).

8. J. S. Miller, *Am. J. Phys.* **27**, 367 (1959).

9. J. Tyndall, *Sound* (3rd ed.), Longman: London (1975).

10. L. H. Cadwell, *Phys. Teach.* **32**, 42 (1994).

11. F. S. Crawford, *Am. J. Phys.* **42**, 278 (1974).

12. D. R. Lapp, *Phys. Teach.* **35**, 314 (1997).

13. S. S. Stevens, *Journ. Acoustical Soc. Am.* **6**, 150 (1935).

14. A. Cohen, *Journ. Acoustical Soc. Am.* **33**, 1363 (1961).

15. W. B. Snow, *Journ. Acoustical Soc. Am.* **8**, 14 (1936).

16. D. Lewis and M. Cowan, *Journ. Acoustical Soc. Am.* **8**, 20 (1936).

17. G. Von Bekesy, *Experiments in Hearing*, McGraw-Hill: New York (1960).

18. P. Davidovits, *Physics in Biology and Medicine* (2_{nd} ed.), Harcourt/Academic Press: San Diego (2001).

Table 3.2

The notes of the equal-tempered scale for the octave above middle C[5]

Note	Frequency (Hz)
C	261.63
C#	277.18
D	293.66
D#	311.13
E	329.63
F	349.23
F#	369.99
G	392.00
G#	415.30
A	440.00
A#	466.16
B	493.88
C	523.25

[5] A-440 is used as the standard, and the ratio between successive notes is $2^{1/12} = 1.05946$.

3.8
Beat Frequencies

Two sound sources of equal amplitudes and nearly equal frequencies exhibit beats if the frequency difference is less than about 10 Hertz.

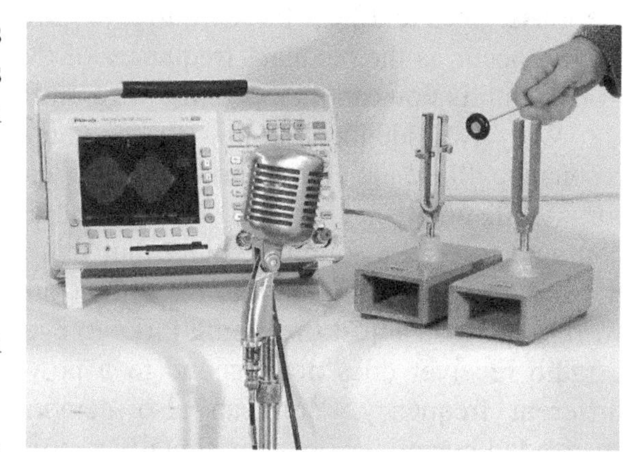

MATERIALS

- two tuning forks or organ pipes with adjustable pitch[1]
- two narrow-neck soda pop bottles (optional)
- two sine-wave generators with amplifier and loudspeaker (optional)
- oscilloscope (optional)
- two glass tubes to fit in the organ pipes (optional)
- Nichrome® wire and power supply (optional)
- overhead projector and transparencies of sine waves (optional)

PROCEDURE

The beating of two tones with nearly equal pitch sounds to the ear like a single tone with their average pitch, whose intensity varies at the beat frequency. You can perform the demonstration using any two sound sources, but the effect is most striking if the sources are sinusoidal and of equal intensities. The cleanest effect results from using two sine-wave generators of equal amplitude connected to the same loudspeaker, in which case the amplitude varies from zero to twice the amplitude of each source [1–4]. You can display the voltage applied to the loudspeaker on an oscilloscope.

Two identical tuning forks also work well if you place a small piece of wax on one of the forks to alter its frequency. You can also use a single tuning fork and walk with it toward a flat wall or blackboard, producing a reflected sound that is Doppler shifted in the opposite direction from the Doppler-shifted sound of the moving tuning fork (see section 3.5) [5]. Similarly, you can use two identical organ pipes if you detune one by partially covering its mouth or by adjusting a movable plug if the pipe is so equipped or

[1] Available from Arbor Scientific, Carolina Biological Supply Company, Frey Scientific, and Sargent-Welch.

by letting some helium gas flow into it (see section 3.3). More simply, blow over the mouth of a pair of soda pop bottles with a nearly equal amount of water in them [6].

You can use the beating of two identical organ pipes to illustrate the variation of the speed of sound with temperature [1]. Place two identical glass tubes, one in each organ pipe, and tune the pipes until the beats disappear. Lower a Nichrome® wire into one of the glass tubes and heat the wire to 250–300°C by an AC current. A noticeable beating should occur as the resonant frequency of the heated pipe rises with the increased speed of the sound. You can then cool the wire until the beats disappear.

With two identical tuning forks, you can also demonstrate the phenomenon of resonance. Place the tuning forks near one another and strike one with a mallet. Let it ring for a while and then grab it in your hand to dampen the sound. The other tuning fork will continue to emit a sound. Explain that this is how a radio receiver responds to a radio transmitter when you tune it to the same (radio) frequency. Show that the effect does not work when you adjust the tuning forks to even slightly different frequencies. That is why a radio receiver does not respond to a powerful nearby transmitter if you tune it to a different frequency. You can also demonstrate resonance and beats using masses suspended by springs on a horizontally supported meter stick [7].

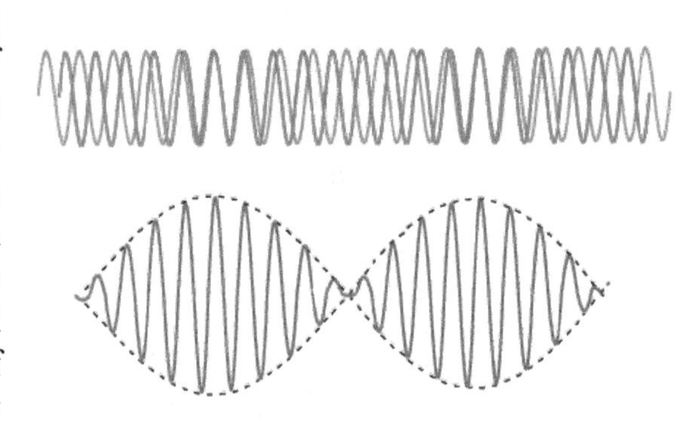

You can explain the beating graphically by drawing two sine waves of slightly different frequencies on the blackboard and showing how they constructively interfere at certain times and destructively interfere at others. You can illustrate the phenomenon more effectively with an overhead projector and two transparencies showing sine waves of slightly different frequencies. Many photocopiers allow you to enlarge or reduce an image.

For an audience versed in trigonometry, explain beats using the relation

$$\sin \omega_1 t + \sin \omega_2 t = 2\sin[(\omega_1 + \omega_2)t/2]\cos[(\omega_1 - \omega_2)t/2]$$

Two sounds of frequency ω_1 and ω_2 are perceived as a single frequency $(\omega_1 + \omega_2)/2$ with an intensity that peaks twice during each cycle of the frequency $(\omega_1 - \omega_2)/2$. Thus the beat frequency is $\omega_1 - \omega_2$ (assuming ω_1 is the larger frequency). If the two sounds are of unequal amplitude, the cancellation will not be perfect when the waves destructively interfere, and the intensity variation of the beats will be less pronounced. Note that there is no wave present at the beat frequency. You can illustrate this fact using two frequencies above the audible range of hearing but adjusted so that the beat frequency is within the audible range. However, if the sounds are sufficiently intense, it is sometimes possible to hear the beat frequency of two inaudible sounds because of nonlinearities in the ear.

DISCUSSION

Beats provide a means for musicians and piano tuners to tune their instruments. A violinist will adjust the tension in a violin string until a given note gives a zero beat frequency with a note played by the concertmaster. Musicians can then adjust other notes so that their overtones beat with the overtones of the standard. In fact, the just diatonic scale has its notes related in frequency by the ratios of small integers to minimize the beating of the overtones when playing two notes together. Such a scale gives a pleasing sound, although musicians more typically tune their instruments to the equal-tempered scale in which the notes have twelve even logarithmic intervals within each octave and thus produce some low-frequency beating among their harmonics (see section 3.7).

Beats appear to play an important role in our perception of the quality of music [8]. A low-frequency beat (a few Hertz or less) can sound pleasing, in that it emulates the trembling of the human voice. Rapid beats can be unpleasant to the ear, just as a flickering light is unpleasant to the eye and a fingernail scratched across a surface is unpleasant to the touch. Thus when one singer in a chorus or one instrument in an orchestra is off key, the resulting beats render the sound highly obnoxious.

The famous German physicist and acoustician (and chronically bad lecturer) Hermann von Helmholtz (1821 – 1894) had a reed organ built that was based on a 24-note scale optimized to reduce the beating between the harmonics of the various notes. The sound reportedly was pleasing but apparently not sufficiently so as to revolutionize the way in which music is written and played. Of course, some instruments, such as a trombone, can produce a continuum of notes.

HAZARDS

There are no precautions other than keeping the sound intensity at a safe level.

REFERENCES

1. H. A. Robinson, Ed., *Lecture Demonstrations in Physics*, American Institute of Physics: New York (1963).
2. R. E. Miers and W. D. C. Moebs, *Am. J. Phys.* **42**, 603 (1974).
3. S. H. Vegors, *Am. J. Phys.* **43**, 1103 (1975).
4. H. F. Meiners, *Physics Demonstration Experiments*, Vol I, The Ronald Press Company: New York (1970).
5. T. B. Greenslade, *Phys. Teach.* **31**, 443 (1993).
6. N. Chareonkul, *Phys. Teach.* **35**, 490 (1997).
7. H. R. Blacksten, *Phys. Teach.* **32**, 554 (1994).
8. H. L. F. Helmholtz, *On the Sensations of Tone as a Physiological Basis for the Theory of Music*, Dover: New York (1954).

3.9
Breaking a Beaker with Sound

A glass beaker exposed to a sufficiently intense sound wave at its natural resonant frequency will shatter.

MATERIALS

- 2-liter, clean, glass beaker or large wine goblet
- Styrofoam® pad
- small rubber mallet, spoon, or Ping-Pong® ball
- small, high-power loudspeaker (such as a horn driver)
- audio amplifier (at least 100 watts)
- variable-frequency, sine-wave generator
- safety glasses
- oscilloscope (optional)
- microphone (optional)
- digital frequency meter (optional)
- strobe light (optional)

PROCEDURE

In this classic demonstration [1–4] connect the input of the audio amplifier to the variable-frequency, sine-wave generator, and connect the output to the loudspeaker. A short resonant pipe connected to the speaker helps to concentrate the sound. Nested telescoping pipes allow you to tune the pipe to the desired resonant frequency. Place the glass beaker on a material such as a Styrofoam® pad that allows it to ring with minimal damping, and position it a few centimeters from the loudspeaker. Be sure the beaker is clean. Tap the beaker near its open mouth with a small rubber mallet (the kind used with tuning forks) or a spoon to ensure that it rings with a slowly decaying tone. With a microphone and oscilloscope, show the audience that the waveform is sinusoidal, and illustrate how to measure the resonant frequency. Do not place the microphone inside the beaker, since it would then respond to the resonance of the air column inside the beaker rather than the resonance of the glass vibration. Alternately, connect the loudspeaker directly to the input of the oscilloscope and use it as a microphone. You must tune the sine-wave generator precisely to the same frequency as the natural resonant frequency of the beaker. A 2-liter beaker will resonate at a frequency of about 500 Hz.

Without an oscilloscope, you can adjust the frequencies to be equal by repeatedly tapping the beaker with the mallet and listening for the beats with the sound from the loudspeaker as you slowly vary the generator frequency. This works best if the sound level is about the same from the loudspeaker and from the beaker. Alternately, put the loudspeaker near the beaker at a low sound level and listen for the increase in sound intensity as you approach the resonant frequency. Another method is to watch or feel the vibrations of the beaker with your fingertips as you vary the frequency. Fold strips of paper over the edge of the beaker to indicate the amplitude of the vibration or place inside the beaker a Ping-Pong® ball, which bounces violently as you approach the resonance. It is best to find the proper frequency beforehand using a digital frequency meter and to record the value (to a precision of at least one Hertz) with a marker pen on the beaker to ensure that you can quickly find the resonance condition for the demonstration, although there is pedagogical value in showing how you find the resonance if time permits.

Caution the audience members to cover their ears, and turn up the sound to a high level (at least 90 dB). A slight readjusting of the frequency may be required. You must adjust the frequency *very* slowly and wait a few seconds for the oscillations to build to a value sufficient for the glass to break.

You can also use a wine goblet or any other glassware with vertical sides that exhibits a high-quality resonance in the acoustical range. All that is required to test the quality of the resonance is to tap the glass with a fingernail or a pencil and listen to how long the resulting ringing lasts. Any glassware that will ring for more than a few seconds should suffice. Glass with a high lead content works best, although expensive glassware is not required. Avoid glasses that are very narrow or very shallow or that have a thick bead of glass around the lip. You can put a split rubber hose over the stem of the glass and support it in any desired position and orientation using a standard rod clamp and ring stand.

In a variation of the demonstration, you can use a fixed-frequency sound source, tuned slightly below the natural resonant frequency of the glass, and then slowly add water or other liquid to the glass until its resonant frequency decreases to match that of the sound source, although too much water will damp the oscillations.

A strobe light tuned near the resonant frequency of the beaker or one of its harmonics makes the vibrations of the glass apparent, although visibility is difficult except for those relatively close. Use a video camera and monitor or LCD projector to improve visibility [5]. Prior to breaking, the glass may deflect as much as a centimeter, which most people find amazing.

DISCUSSION

This demonstration illustrates that sound is a compressional wave with a characteristic frequency. The wave carries significant energy from the source to the point where it is absorbed. More importantly, the demonstration illustrates the phenomenon of resonance. The Q (quality) of the resonance is 2π times the number of cycles (the number of radians) required for the energy of an oscillation to decay to 1/e (~37%) of its initial value. Thus the highest Q resonance is the one for which the ringing lasts the longest. A piece of glassware with a resonance at 1,000 Hz that decays to about 37% of its initial energy in one second has a Q of about 6,000. The Q is also a measure of the amplitude that an oscillator driven at its resonant frequency will acquire. Thus the glass with a Q of 6,000 will vibrate with about 6,000 times more energy when driven at resonance than when driven well off resonance with a sound of the same intensity. However, it takes about Q radians (one second in this example) for the oscillation to build to near its full amplitude. You must also match the driving frequency and the natural resonant frequency to a precision of about $1/Q$ (one part in about 6,000 in this example). Since the sound intensity has to be high (in excess of 90 dB), precisely of the right frequency, and sustained for a few seconds, there is little danger of breaking your wine glass or eye glasses while casually listening to music on your stereo!

Dorothy Caruso denied that her husband Enrico (1873 – 1921) could shatter wine glasses with his unamplified voice [6], although rumors persist that he and the Italian opera singer Beniamino Gigli (1890 – 1957) did this. The Memtek company in Fort Worth, Texas, produced a television commercial showing Ella Fitzgerald (1917 – 1996) and others breaking glasses with recordings of their voices on Memorex® audiotape. In 2005, Jaime Vendera shattered a wine glass on the MythBusters TV show.

HAZARDS

Unless you design this demonstration carefully, the sound intensity required to break a beaker can damage one's hearing. Turn up the volume slowly to allow the audience to take protective action. Encourage the audience members to cover their ears. When the glass shatters, the glass fragments tend not to be ejected with great force but rather to fall downward. Even so, it is best to do the demonstration some distance from the audience or to use a transparent protective screen and to wear safety glasses. A transparent box also reduces the sound level outside the box, while increasing it inside. A tray underneath the glass will catch most of the glass fragments, but you should promptly collect the broken pieces and dispose of them, especially if they fall onto the floor.

REFERENCES

1. W. C. Walker, *Phys. Teach.* **15**, 294 (1977).
2. G. P. Jones and W. P. Gordon, *Am. J. Phys.* **47**, 828 (1979).
3. H. Kruglak, R. Hiltbrand, and D. Kangas, *Phys. Teach.* **28**, 418 (1990).
4. B. Berner, *Phys. Teach.* **38**, 269 (2000).
5. W. Rueckner, D. Goodale, D. Rosenberg, S. Steel, and D. Tavilla, *Am. J. Phys.* **61**, 184 (1993).
6. H. Kruglak and R. Pittet, *Phys. Teach.* **17**, 49 (1979).

3.10 Ultrasound

Various sources of sound with frequencies above the range of audibility illustrate the distinction between a physical sound wave and the perception of sound.

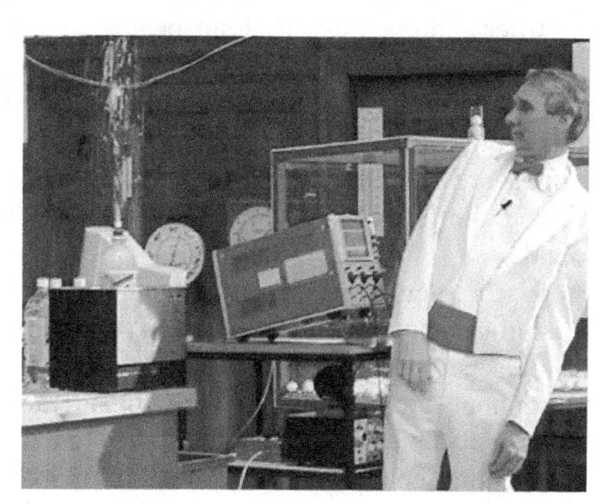

MATERIALS

- ultrasonic dog whistle[1]
- variable-frequency audio oscillator with loudspeaker (optional)
- ultrasonic cleaner (optional)[2]
- carbonated beverage (optional)
- ultrasonic microphone and oscilloscope (optional)

PROCEDURE

You can introduce this demonstration with the old question of whether a tree falling in a deserted forest makes a sound. The answer depends on whether you define sound as a physical wave or the perception of that wave by the ear and brain of a living creature. Sources of ultrasound provide a means to produce a sound wave that humans do not perceive as a sound [1–3]. Ask the audience if there are sounds that no one can hear.

Probably the least expensive and most convenient source of ultrasound is a dog whistle. A dog, trained to come to the sound of the whistle that interrupts the presentation when you first demonstrate it, will get the attention of the audience, or

[1] Available from most pet supply stores.
[2] Available from American 3B Scientific, Carolina Biological Supply Company, Cole Parmer, and Sargent-Welch.

someone can toss you a stuffed dog or make barking sounds. You can usually adjust dog whistles from a frequency that is barely audible to most people to well above the range of hearing. Ultrasonic pest repellers produce an intense burst of ultrasound every few minutes.

You can use a variable-frequency audio oscillator with a high-fidelity audio amplifier and good quality loudspeaker (tweeter preferred) to explore the whole range of audible frequencies. You can point out that high-frequency audible response deteriorates with age (a medical condition known as "presbycusis"). If the audience consists of many ages, you could perform an experiment on this. One reason that children and small animals are more responsive to high frequencies is a simple matter of size scaling. An ear with smaller structures is more sensitive to the shorter wavelengths of a high-frequency sound.

Use a microphone connected to an oscilloscope to prove that there is a sound present. Most microphones have low sensitivity to ultrasonic frequencies, but can usually indicate the presence of a sound with a sensitive oscilloscope. A high-pass filter between the microphone and oscilloscope is of considerable help. A small capacitor will often suffice. You can also purchase special ultrasonic microphones.

Another common source of sound above the audible range is an ultrasonic cleaner, which usually consists of a vat of liquid with an ultrasonic transducer that creates sound waves in the liquid. Typically the liquid is some organic compound capable of dissolving oil and grease, but water is adequate for this demonstration. Jewelers often use ultrasonic cleaners to clean rings and other jewelry. A freshly opened plastic bottle of carbonated beverage partially immersed in the liquid will erupt and spray its contents several feet into the air especially if equipped with a nozzle of some sort. Explain that the reason is that the ultrasonic waves shake the liquid rapidly just like most people have done manually with a bottled carbonated beverage. Ask what the gas is, or point out that it is carbon dioxide. A carbonated beverage without sugar or other added ingredients (seltzer) will prevent making a sticky mess.

Because of the short wavelengths, ultrasonic transmitters and receivers are well suited for the display of interference and diffraction effects usually demonstrated with microwaves or light. This helps dispel the mystery of why, if light and sound are both waves, we can hear around a corner but we cannot see around one. The short wavelength makes ultrasound useful for medical imaging, and it is safer than X-rays, especially for imaging fetuses. The technique is the sonic equivalent of radar, which uses short-wavelength electromagnetic waves to image objects by their reflection. When used in combination with the Doppler effect (see section 3.5), ultrasound can be used to measure the speed of blood flow in the body. Intense ultrasound is also useful for local heating of the body (diathermy) to relieve pain and heal injuries, and even to destroy tissue. It is used routinely for destroying kidney stones and gallstones (lithotripsy).

DISCUSSION

A sound wave is a compressional disturbance that will propagate in most materials. The range of human hearing is usually about 20 Hz to 20 kHz, although it is more accurate to say that the ear is most sensitive to frequencies around 4,000 Hz and becomes progressively less sensitive at higher and lower frequencies, requiring greater sound intensity to produce a detectable response. The product of the frequency and the wavelength is equal to the speed of sound in the medium. In air at 20°C, the speed of sound is 343 m/s, and so the corresponding wavelengths are 17.15 m to 1.715 cm, respectively. Typical ultrasonic sources produce frequencies from about 20 to 100 kHz, corresponding to wavelengths as short as a few millimeters. Special electronically driven (piezoelectric) crystals can produce ultrasound frequencies up to several megahertz.

The reason it is difficult both to produce and to detect ultrasound is that something has to move to create or to respond to the sound wave. The inertia of the object (eardrum or whatever) prevents it from moving appreciably during the short period of the wave. Even so, the human ear is a remarkable instrument, responding to displacements of the eardrum that are hardly larger than the diameter of an atom!

HAZARDS

It is possible to damage one's hearing by exposure to sufficiently intense ultrasound even though the sound is inaudible, but this is unlikely unless you place an ultrasonic transducer directly in the ear, which is definitely not recommended. Internal damage can occur if one places a finger or other part of the body in the vat of an ultrasonic cleaner with the sound turned on.

REFERENCES

1. J. A. Zagzebski, *Essentials of Ultrasound Physics*, Mosby: Saint Louis (1996).

2. E. Papadakis, *Ultrasonic Instruments and Devices*, Academic Press: San Diego (2000).

3. D. Cheeke, *Fundamentals and Applications of Ultrasonic Waves*, CRC Press: Boca Raton, FL (2002).

4. J. L. Rose, *Ultrasonic Waves in Solid Media*, Cambridge University Press: Cambridge (2004).

4
Electricity

Although electricity has many parallels with motion, it involves a fundamentally different force. Electricity is the study of the motion and effect of electrical charges, just as mechanics is the study of the motion and effect of masses. The subject of electricity acquired a firm theoretical basis through the work of the Scottish mathematician and physicist James Clerk Maxwell[1] (1831 – 1879), whose four equations encompass all electromagnetic phenomena. When Einstein, a generation later, upset much of classical physics, Maxwell's equations survived untouched. Electrical demonstrations, especially those involving high voltages, constitute some of the most spectacular in the whole of physics.

[1] James Clerk (pronounced like "Clark") Maxwell was rather dull in school and was given the name "Dafty" because of his shyness and the unusual clothes designed by his father.

Safety Considerations with Electricity

The use of electricity in demonstrations entails special hazards from electrical burns and electrocution [1]. The hazards of electricity (see table 4.1) depend on the amount of electrical current, its frequency, its duration, its path through the body, and the physical condition of the person exposed to the hazard. Alternating current at 50 or 60 Hz is slightly more dangerous than direct current, but high-frequency currents (greater than 100 kHz) are safer because the muscles and nerves are not sensitive to such frequencies. Even higher frequencies (greater than about 10 MHz) are relatively safe because the currents tend to flow on the surface of the skin, although burns can result from local heating. The resistance of the body varies from about 300 ohms to about 100,000 ohms, and thus even low voltages can produce lethal shocks, especially if the skin is wet at the point of contact. The hand-to-hand resistance is typically 1,000 to 2,000 ohms. Fatalities have occurred at voltages as low as 24 volts. For currents that exceed the "let-go" current of 10–20 mA [2] the person becomes frozen to the circuit, and the current typically rises to a level of about 25 mA where muscular contractions onset. Then the person is either thrown clear of the circuit or the current continues to rise until ventricular fibrillation or cardiac arrest occurs at 50–200 mA, assuming the path of the current is through the heart. Curiously, a larger current of the order of 10 A is often less dangerous because the heart may resume its normal rhythmic activity when the current is removed, as occurs with a defibrillator [3].

Pulsed currents such as one might encounter with the discharge of a Van de Graaff generator or other charged capacitance present special considerations. One can endure currents that would otherwise be lethal if the duration is short enough. For pulses of less than a few seconds' duration, the relevant quantity is the square of the current integrated over the time of the pulse. Values of I^2t greater than about 0.01 A^2sec can cause electrocution for a typical adult. For a reasonably low body resistance of 2,000 ohms, this value implies an energy of about ten joules. Severe shocks can occur at levels ten to a hundred times lower and can startle one into an accident, since it is natural to jerk away from such a shock. Defibrillators typically use a capacitor charged to about 6,000 V with a stored energy of about 200 J [3].

The usual safety precautions entail some combination of the following: copious and redundant insulation, grounding of all exposed metal parts, interlocks, isolation transformers, and ground fault interrupters, which interrupt the current if even a tiny fraction of it returns to ground rather than through the paired conductor. Good practice entails standing on an insulated surface and keeping one hand in your pocket or behind your back while working around high voltage. In performing electrical demonstrations, you should ideally have a knowledgeable assistant trained in cardiopulmonary resuscitation (CPR) always present. The probability of resuscitating someone is good if CPR begins within three minutes but becomes poor after about six minutes. Many groups such as the American Red Cross and the American Heart Association offer CPR classes.

In addition to the danger of electrocution, high electrical voltages can produce corona, which is an electrical breakdown of the air in a region of high electric field. The resulting free electrons can cause the formation of toxic gases such as ozone (O_3), which can corrode the lungs if breathed over too long a period, resulting in pulmonary edema.

Ozone is easily detected by its pungent odor. OSHA (Occupational Safety and Health Administration) has established 0.1 parts per million (PPM) by volume of air (or 0.2 mg/m^3) as the maximum allowable safe concentration of ozone for an eight-hour industrial exposure. Considerably greater concentrations are safe if the exposure is of short duration. Halon® gas fire extinguishers are most suitable for electrical fires.[1]

REFERENCES

1. E. A. Lacy, *Handbook of Electronic Safety Procedures*, Prentice-Hall: Englewood Cliffs, NJ (1977).

2. C. F. Dalziel and W. R. Lee, *IEEE Spectrum* **6**, 44 (Feb 1969).

3. P. Davidovits, *Physics in Biology and Medicine* (2nd ed.), Harcourt/Academic Press: San Diego (2001).

Table 4.1

Average effects of continuous AC or DC electrical currents on healthy adults

Electrical current	Biological effect
1 mA	Threshold of feeling
5 mA	Onset of pain
10–20 mA	Voluntary let-go of circuit impossible
50–200 mA	Ventricular fibrillation or cardiac arrest

[1] Halon® (bromotrifluoromethane) is a compound of carbon with bromine and fluorine ($CBrF_3$), which, like carbon dioxide, is stored in a fire extinguisher as a liquefied, compressed gas that leaves no residue and is safe for human exposure, but unlike carbon dioxide does not displace air from the area where it is dispensed. Because it is a chlorofluorocarbon (CFC), new Halon® production in the United States ceased in 1994, although it is still available in recycled form and is gradually being replaced by more environmentally friendly compounds such as Halotron®.

4.1
Wimshurst Electrostatic Generator

A Wimshurst electrostatic generator produces high voltages at moderate currents to illustrate many principles of electrostatics.

MATERIALS

- Wimshurst electrostatic generator with accessories[1]

PROCEDURE

A simple device capable of producing moderately high voltages for electrostatic demonstrations is the Wimshurst generator [1], invented in 1883 by James Wimshurst (1832 – 1903). Other similar devices were invented by the German physicists Toepler (in 1865), Holtz and Poggendorff (in 1869), Musaeus (in 1871), Voss (in 1880), and many others. More recently the American A. D. Moore performed interesting lecture demonstrations with his Dirod generator [2]. These devices illustrate the conversion of mechanical energy into electrical energy.

In addition to making sparks jump through the air, these devices are capable of operating X-ray tubes, cathode ray tubes, and most types of vacuum and gaseous discharge tubes. You can use them to deflect the flame of a candle burning near a pointed electrode at high voltage or between the plates of a charged capacitor [3]. You can ignite a dust or ethanol vapor explosion with the spark from these machines (see section 2.23). You can use them to drive an electrostatic motor or a pinwheel with corona or to ring a set of chimes consisting of a metal ball suspended by an insulated thread between two bells at opposite electric potential or to separate a mixture of salt and pepper or to precipitate smoke from the air [4–9]. You can touch a fluorescent tube alternately and repeatedly to the electrodes, producing a flash each time. The same vendors who supply Wimshurst generators also offer

[1] Available from American 3B Scientific, Carolina Biological Supply Company, Fisher Science Education, Frey Scientific, Sargent-Welch, Ward's Science, and Scientifics.

accessories for such demonstrations, although you can construct many accessories from inexpensive, readily available components [10]. The spark from the generator will produce ozone (O_3), which is evident by its distinctive odor.

DISCUSSION

These devices work by means of rotating plastic or glass plates that become electrically charged and transfer their charge to a set of external combs, which typically charge a pair of Leyden jar capacitors. The disks rotate by a hand crank connected to the disks with an insulated belt. A useful improvement is to add a motor drive with variable speed control. The capacitors discharge when a spark occurs between a pair of metal balls whose proximity is adjustable. The distance over which a spark will jump between two spherical balls provides a reasonably accurate measure of the voltage (see table 4.2). Although these devices produce lower voltage than Van de Graaff generators, they can briefly produce moderately large currents. In most cases you can remove the Leyden jars from the circuit to illustrate the difference between a high-current and a low-current discharge of similar voltage. The Leyden jar ("Leiden" in Dutch) is the forerunner of the modern-day capacitor and was invented in 1746 at the University of Leyden (the oldest and most famous university in Holland) by Pieter Van Musschenbroeck (1692 – 1761), although the effect had also been observed by Ewald Georg von Kleist (ca. 1700 – 1748) in 1745 in Poland [11, 12]. In old naval manuals, the unit of capacitance was a "Jar."

HAZARDS

Most commercially available Wimshurst generators are designed so that a nonlethal amount of electrical energy is stored in the Leyden jars. However, a discharge to the body is surely painful, and might startle one into an accident. If you connect significant additional capacitance to the output of the Wimshurst generator, the hazard can increase substantially. The spark can also cause burns and ignite flammable or volatile materials nearby. Ozone is a toxic gas that can corrode the lungs if breathed over too long a period, resulting in pulmonary edema. Be sure to short out the electrodes when you have completed the demonstration.

REFERENCES

1. G. B. Putz, *Popular Electronics* **28**, 29 (Dec 1990).

2. A. D. Moore, *Electrostatics: Exploring, Controlling, and Using Static Electricity*, Doubleday: Garden City, NY (1968).

3. M. Robinson, *Am. J. Phys.* **30**, 366 (1962).

4. M. Robinson, *Bibliography of Electrostatic Precipitator Literature*, Southern Research Institute: Birmingham, AL (1969).

5. F. T. Cameron, *Cottrell, Samaritan of Science*, Doubleday: Garden City, NY (1952).

6. H. J. White, *Industrial Electrostatic Precipitation*, Addison-Wesley: Reading, MA (1963).

7. A. D. Moore, Ed., *Electrostatics and its Applications*, John Wiley & Sons: New York (1973).

8. J. Bohm, *Electrostatic Precipitators*, Elsevier Scientific Publishing Company: New York (1982).

9. J. S. Miller, *Physics Fun and Demonstrations*, Central Scientific Company: Chicago (1974).

10. W. R. Mellen, *Phys. Teach.* **27**, 86 (1989).

11. F. Cajori, *A History of Physics*, Dover: New York (1962).

12. T. B. Greenslade, *Phys. Teach.* **32**, 536 (1994).

Table 4.2

Approximate gap spacing (in cm) for electrical breakdown between two identical spheres of various diameters [1]

kV	2.5 cm	3.0 cm	4.0 cm	5.0 cm	10.0 cm
10	0.30	0.30	0.30	0.30	0.30
20	0.61	0.61	0.61	0.61	0.61
30	0.95	0.95	0.95	0.95	0.95
40	1.40	1.32	1.30	1.30	1.30
50	2.00	1.82	1.73	1.71	1.65
60	2.81	2.40	2.21	2.16	2.02
70	4.05	3.16	2.80	2.68	2.41
80		4.40	3.50	3.26	2.82
90			4.40	3.93	3.28
100				4.76	3.75

4.2
Van de Graaff Generator

A Van de Graaff generator illustrates many principles of high-voltage electrostatics, such as making a person's hair stand on end.

MATERIALS

- Van de Graaff generator[1]
- grounded conducting sphere
- grounded pointed needle
- insulated stool
- hand mirror or camera (optional)
- puffed rice or Styrofoam® peanuts (optional)
- cotton ball or silvered balloon (optional)
- soap bubbles (optional)
- fur and bright light (optional)
- stack of aluminum cups (optional)
- fluorescent tube (optional)

PROCEDURE

A standard physics lecture demonstration is the Van de Graaff generator, named after the American physicist Robert J. Van de Graaff (1901 – 1967), who developed it in the 1930s. You can make a Van de Graaff generator with a hollow metal dome, a belt to transfer charge to the dome, and a motor, preferably with variable speed control [1–7]. More commonly, you can purchase Van de Graaff generators from vendors of scientific demonstration apparatus. The limiting voltage is determined by the electrical breakdown of air (about 30 kV per centimeter at atmospheric pressure), and so a Van de Graaff

---

[1] Available from American 3B Scientific, Arbor Scientific, Carolina Biological Supply Company, Fisher Science Education, Frey Scientific, Klinger Educational Products, PASCO Scientific, Sargent-Welch, Science First, Ward's Science, and Scientifics.

generator with a 10-cm-radius dome is capable of producing up to about 300 kV. The voltage may be considerably less if the humidity is high or if sharply pointed, grounded conductors are nearby allowing corona [8] to discharge the dome. You should clean the dome and insulated supporting column before use. Despite common belief, the electrical breakdown of air is only slightly affected by humidity, but surface leakage current is greatly enhanced by moisture, limiting the voltage that can be obtained.

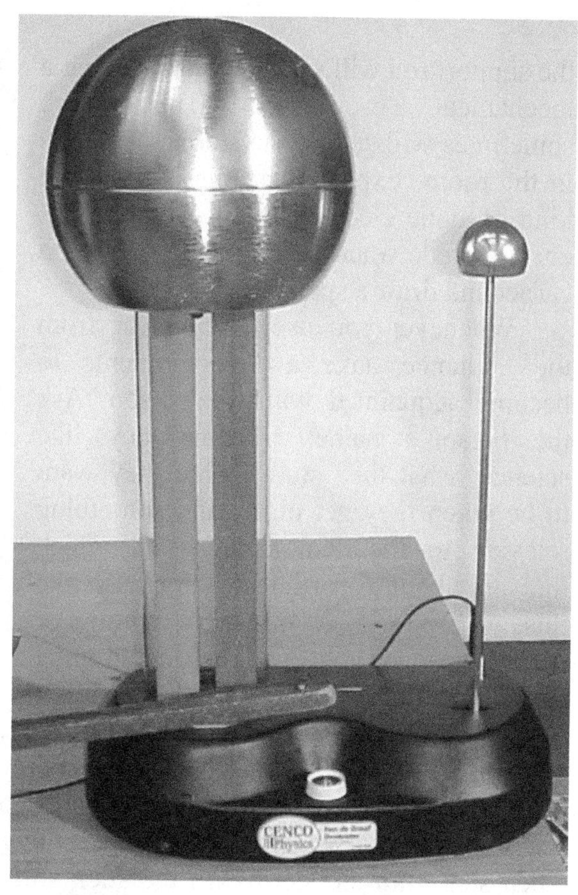

The standard demonstration involves having a volunteer stand on an insulated stool (of a height greater than the diameter of the dome to ensure adequate insulation) and place a hand on the dome while the Van de Graaff generator is turned on. Offer the volunteer a "hair-raising experience." After a moment, the person's hair will begin to stand on end. A hand mirror is useful to allow the volunteer to see what is happening (but do not hand it to someone charged to high voltage!). A digital photograph, perhaps on the volunteer's cell phone, also makes a good souvenir.

It is best to choose someone with medium length, fine, dry hair. Shaking the head helps make the hair stand up, and having the volunteer look in different directions enables the audience to see better. The person should be well away from any grounded conductors and told to leave the hand on the dome at all times until told to remove it. The volunteer should remove any sharply pointed jewelry, which tends to limit the voltage obtained. When you finish, turn off the motor, and slowly discharge the dome by bringing a grounded, pointed needle near it while the person is still in contact with it. The hair will drop back into place as corona from the needle discharges the dome, and the volunteer can then step down from the stool. If the stool is small or at all tipsy, help the person get on and off the stool. You should clean the stool before use.

Pick up the stool and explain its purpose to the audience. Explain that rubber shoes would not suffice because they are too thin to withstand the very high voltage used here, although they provide some protection at lower voltages. Mention that it is not a good idea to be in bare feet while working around electricity. Warn people that if they are ever outdoors in a lightning storm and their hair begins to stand up, that they should take cover because there is a chance that lightning is about the strike them.

Afterwards, you could show the distance over which a spark will jump by bringing a grounded sphere near the dome. The increased capacitance of the sphere makes the spark more intense and visible. If the grounded sphere is mounted on a flexible rod, the attractive force due to the induced charge in the sphere will make the support rod bend toward the dome. When a spark occurs, neutralizing the charge on the grounded sphere,

the support rod will straighten, producing a mechanical oscillation. The spark sometimes will set off the smoke detectors in the room. Explain that this is the same kind of static electricity that everyone has encountered when they walk across a carpet and draw a spark to a doorknob.

Whenever you use a volunteer from the audience, take a few moments to become acquainted with the person. Ask the person's name, whether they like science, what they study, what they want to be when they get older, and something relevant to the demonstration in which they are about to participate. In a case like this, ask if they have electricity at home or if they have ever been out in a lightning storm. Ask if they are scared of electricity. You could also ask if they have a cat or dog and whether they have ever made sparks by rubbing the animal. To get a laugh, ask if they are a lawyer.

If you prefer to do the demonstration yourself rather than with a volunteer, turn on the Van de Graaff generator and touch the grounding stick to the dome with one hand. Then step onto the stool with the stick still in contact with the dome. Touch the dome with your other hand and then remove the stick. When you are done, bring the grounding stick up to the dome again while keeping your hand on the dome. Then step down from the stool, and turn off the generator with the grounding stick still in contact with the dome. You can draw a spark to your hand from the Van de Graaff generator without too much discomfort if you make a fist and draw the spark to your knuckles, where there are few pain sensors.

You can also do other demonstrations with the Van de Graaff generator. Puffed rice or Styrofoam® peanuts placed in a container on the dome will fly off in all directions when you turn the generator on. A pinwheel placed on the dome will rotate from the corona given off its points. A fluffy cotton ball or silvered balloon [9−11] released near the generator will be attracted to the dome and then repelled after it touches the dome and becomes charged by the dome. Someone at ground potential can blow soap bubbles toward the dome, or someone on the stool can blow them away from the dome while touching the dome [12, 13]. You can toss bits of fur into the region between a charged and grounded dome and illuminate it with a strong light to illustrate qualitatively the shape of the electric field [14]. A stack of aluminum cups placed on the dome will jump off one-by-one. In a darkened room, you should be able to show that a fluorescent tube held near the dome glows dimly.

DISCUSSION

The Van de Graaff generator demonstrates a number of principles, foremost of which is the relationship between charge and voltage, $Q = CV$, where C is the capacitance of the dome. The capacitance of a sphere with respect to infinity is about 1.1 pF per centimeter of radius. The electrical energy stored in the Van de Graaff generator is $CV^2/2$. Energies below about 10 joules (corresponding to a dome with a 27-cm radius) are relatively safe. The volunteer is also safe because the insulated stool prevents significant current from passing though the body. Hair stands up because it acquires charges of like sign that tend to repel one another. The Van de Graaff generator illustrates static electricity, with which most people are familiar. It is static in the sense that the electrical charges do not flow from place to place but simply reside on the dome in contrast to the electricity you have in your home, which flows through the appliance plugged into the socket. A Van de Graaff generator with a hand crank allows you to dispel the erroneous belief that the electricity is coming from the power lines since the dome will charge even when the device is not connected to a source of electricity.

The Van de Graaff generator was one of the earliest particle accelerators used for nuclear physics research. Modern Van de Graaff generators and other similar electrostatic accelerators still serve that purpose and are capable of voltages in excess of 20 million volts [15].

HAZARDS

Small Van de Graaff generators (less than a few hundred thousand volts) are relatively safe since the charging current is typically only a few microamperes and the capacitance is too small to allow much of a charge buildup, provided you do not add external capacitance [16]. Even so, it is probably best not to let someone with a weak heart or a heart pacemaker touch the dome. The greater danger is that a spark to the person will be startling and might induce recoil that could cause injury, especially if the volunteer falls from the stool.

REFERENCES

1. R. J. Van de Graaff, J. G. Trump, and W. W. Buechner, *Rep. Progr. Phys.* **11**, 1 (1948).
2. A. W. Simon, *Am. J. Phys.* **22**, 318 (1952).
3. H. Walton, *Popular Science* **185**, 142 (1964).
4. C. L. Strong, *Scientific American* **225**, 106 (Aug 1971).
5. A. D. Moore, Ed., *Electrostatics and its Applications*, John Wiley & Sons: New York (1973).
6. A. W. Simon, *Am. J. Phys.* **43**, 1108 (1975).

7. R. E. Berg, *Phys. Teach.* **28**, 281 (1990).

8. L. B. Loeb, *Electrical Coronas*, University of California Press: Berkeley, CA (1965).

9. P. Hood, *Am. J. Phys*. **14**, 445 (1946).

10. A. V. Baez, *Am. J. Phys*. **25**, 301 (1957).

11. M. D. Daybell and R. J. Liefeld, *Am. J. Phys*. **31**, 135 (1963).

12. G. R. Gore, *Science Teacher* **39**, 48 (1972).

13. R. Prigo, *Am. J. Phys*. **44**, 606 (1976).

14. J. L. Smith, *Phys. Teach.* **27**, 358 (1989).

15. P. H. Rose and A. B. Wittkower, *Scientific American* **223**, 24 (Aug 1970).

16. A. D. Moore, *Electrostatics: Exploring, Controlling & Using Static Electricity*, Doubleday: Garden City, NY (1968).

4.3 Electrophorus

A static electric charge on an insulator can repeatedly induce a charge in a conducting plate, raising the voltage to a high value and making sparks.

MATERIALS

- flat plastic insulator (preferably Teflon®)
- flat, smooth, round conducting plate with rounded edges, approximately 20 cm in diameter but smaller than the insulator, with an insulated handle
- cat's (or rabbit's) fur or cloth (preferably wool or silk)
- electroscope (optional)
- small neon bulb or fluorescent tube (optional)

PROCEDURE

Place the flat plastic insulator (once known as the "cake") on a grounded surface, and rub the upper surface of the insulator with the cat's fur or cloth to deposit an electrical charge on it by triboelectrofication [1] (see table 4.3). Comment that the cat donated its fur after it no longer needed it. You will know that it is charged by the tingling sensation (called "formication") you feel when you bring your hand near the insulator. Lower the conductor onto the insulator and touch its upper surface with your finger to discharge it, drawing a spark. Then raise the conducting plate off the insulator with the insulated handle, being careful not to touch the plate or even coming too close to it. You can then draw a spark of a centimeter or more in length to the edge of the conducting plate with your finger tip [2]. Lifting the conductor higher gives a larger spark [3], but waiting too long after lifting the plate will allow some of the charge to bleed off,

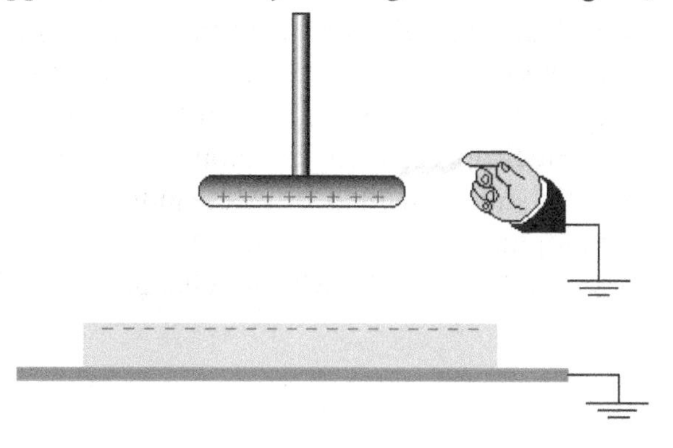

reducing the spark. You can feel and hear the spark, but you may need to reduce the room light to see it. You can repeat as often as desired without recharging the insulator [4] since very little of the negative charge transfers to the conductor. Offer to let someone in the front row touch it, but ask the volunteer if he or she is a lawyer before doing so.

Optionally, touch the conducting plate to an electroscope to measure the charge on it after raising it off the insulator, or use it to light a small neon bulb or fluorescent tube momentarily [5]. You can determine the sign of the charge on the plate by seeing which electrode of the neon bulb illuminates, since the glow will be around the more negative electrode. You can use the spark for vapor ignition demonstrations such as the ethanol vapor explosion in section 2.23.

You can construct an inexpensive version of the electrophorus demonstration using a Styrofoam® dinner plate or an old LP record for the insulator, and a disposable aluminum pie pan with a Styrofoam® drinking cup glued to it as the insulating handle [4, 6]. Most household glues will not work because they will dissolve the Styrofoam®. An aluminum or stainless steel frying pan with a flat bottom and insulated handle can substitute for the pie pan.

DISCUSSION

The insulator forms a capacitor between the underneath ground and the conducting plate on its top. The sign of the charge on the upper surface of the insulator will depend on what insulator you use and what material you rub against it. Assume the charge is negative. When you place the initially uncharged conductor on top of the insulator, the negative charges on the insulator repel the negative charges in the conductor, which then move to the top of the conducting plate, leaving a positive charge on the bottom. When you touch the top of the plate, the negative charges conduct to ground, leaving a net positive charge on the plate. The plate thus charges by induction.

When you raise the plate off the insulator, the charge Q on it remains the same and redistributes throughout the surface of the conductor, but the capacitance C decreases by a large factor. Since $V = Q/C$ for a capacitor, the voltage V rises to a large value, often reaching as much as 50 kV, which can produce a spark of more than a centimeter. The electrostatic energy $Q^2/2C$ rises by a similar factor, with the energy coming from the work you do when lifting the conducting plate against the Coulomb force attracting the plate to the insulator. The force can be so great that the insulator may remain stuck to the conductor. Since the charges do not actually leave the insulator, you can repeat the demonstration many times without recharging the insulator. In fact, keeping the plate on top of the insulator prevents the charges on the insulator from leaking off, much as the keeper on a permanent horseshoe magnet tends to maintain the magnetic field in the magnet. The charge can remain for many months [7].

Electrophorus was developed in 1762 by Johan Wilcke (1732 – 1796) and perfected in 1775 and so named by Alessandro Volta (1745 – 1827), who was also responsible for developing the battery and after whom the unit of the volt was named. The term “electrophorus” is Greek for “charge carrier.”

HAZARDS

The spark is easily felt and might be startling the first time or two, but the charge is small enough not to pose a significant hazard. If you use the electrophorus to charge a Leyden jar, the resulting charge can be much larger and potentially dangerous. It is best not to do this demonstration with someone wearing a heart pacemaker.

REFERENCES

1. D. S. Ainslie, *Am. J. Phys.* **35**, 535 (1967).
2. R. M. Sutton, *Demonstration Experiments in Physics*, McGraw-Hill: New York (1938).
3. D. R. Lapp, *Phys. Teach.* **30**, 454 (1992).
4. R. A. Morse, *Teaching about Electrostatics: An AAPT/PTRA-plus Workshop Manual*, American Association of Physics Teachers: College Park, MD (1992).
5. W. Layton, *Phys. Teach.* **29**, 50 (1991).
6. A. Van Heuvelen, L. Allen, and P. Mihas, *Phys. Teach.* **37**, 482 (1999).
7. D. S. Ainslie, *Phys. Teach.* **20**, 254 (1982).

Table 4.3

Triboelectric series (approximate)

Material[1]
Human skin
Asbestos
Rabbit fur
Glass
Human hair
Mica
Nylon®
Wool
Lead
Cat fur
Silk
Aluminum
Paper
Cotton
Steel
Wood
Lucite®
Sealing wax
Amber
Polystyrene
Rubber
Nickel, Copper
Brass, Silver
Gold, Platinum
Sulfur
Acetate, Rayon®
Polyester
Celluloid®
Polyurethane
Polyethylene
Polypropylene
PVC (vinyl)
Silicone rubber
Teflon®

[1] Objects near the top of the list tend to acquire positive charge, and objects near the bottom of the list tend to acquire negative charge when rubbed against one another.

4.4
Exploding Wire

A thin wire or strip of aluminum foil vaporizes when a large capacitor discharges through it.

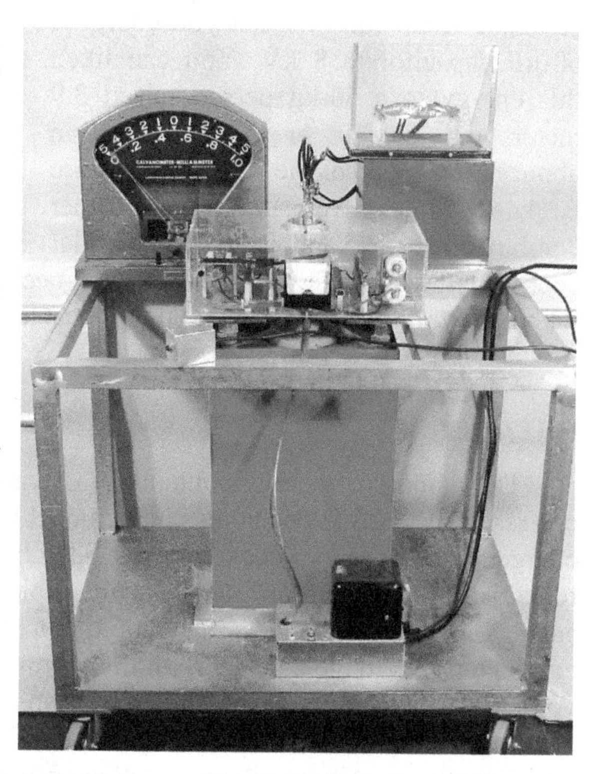

MATERIALS

- 60-μF, 10-kV capacitor with charging circuit[1]
- class-A ignitron or spark gap switch
- thin copper wire (#22 gauge) or strip of aluminum foil
- large voltmeter, 10-kV full scale (optional)
- safety glasses

PROCEDURE

Explain the operation of a capacitor to the audience. Liken it to an automobile battery, but with much higher voltage. Charge the capacitor to a good fraction of its rated voltage, and then discharge it with an ignitron or spark gap switch into a thin wire or strip of aluminum foil, about 20 cm long, which instantly vaporizes. Warn the audience to cover their ears, since the resulting noise can be quite loud. Place a protective piece of Plexiglas® between the wire and the audience, wear safety glasses, and stand back a safe distance.

DISCUSSION

This demonstration exhibits in a dramatic way the fact that electricity is a form of energy. In the process, the electrical energy changes into five other forms: motion, heat, sound, magnetism, and light. The wire or foil usually completely disappears, although it is doubtful that it all vaporizes. More likely, a portion vaporizes and the remainder is flung into some distant corner of the room by the magnetic forces. You can see that some of the foil vaporizes because it eventually coats the protective Plexiglas®, requiring that you occasionally clean it with soapy water or window cleaner. It demonstrates phase

---

[1] The same apparatus can be used in the Can Crusher demonstration (see section 5.4).

changes, since the solid aluminum foil changes into liquid, gas, and plasma, but it returns to the solid form when it cools and condenses onto the Plexiglas®.

The stored electrical energy is $CV^2/2$ or about 1,920 joules when you charge a 60-μF capacitor to 8 kV. You can liken this energy to a 50-kg person raised 3.9 meters into the air, or to the energy in about half a food calorie (one food calorie is 4,186 joules). Capacitors with other combinations of C and V giving a similar energy will also work, but the capacitor needs to be a type designed for energy storage with high voltage and low internal inductance and resistance.

HAZARDS

The amount of energy contained in a capacitor of this size is deadly. Electrocution can result instantly if one comes into contact with the terminals of the capacitor while it is charged. The wire explodes with considerable force and noise. Some voltage generally remains on the capacitor after the wire explodes. A bleeder resistor in parallel with the capacitor is recommended but should not be relied upon to discharge the capacitor completely. Store the capacitor with its terminals shorted. You should practice at reduced voltages, and take the necessary precautions to prevent the hot wire or foil from being ejected toward the audience. Warn the audience of the impending noise. Capacitors occasionally short internally and explode especially if fully charged to their rated voltage. The capacitor should not be in direct line-of-sight of the audience. Interlock the charging supply so that a curious spectator cannot initiate a charge. Wear safety glasses for this demonstration.

4.5 Jacob's Ladder

An electrical discharge occurs with a high-voltage power supply connected to a pair of conducting bars close together at the bottom and farther apart at the top, between which a discharge strikes at the bottom and rises to the top and then strikes again at the bottom.

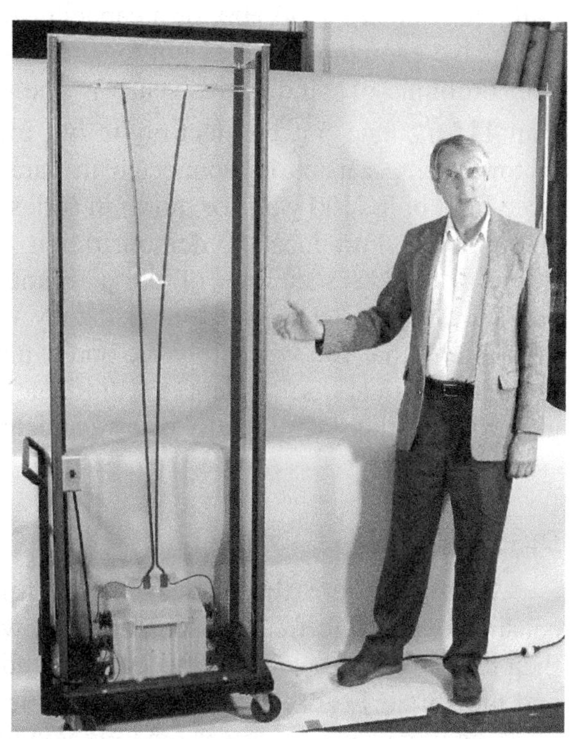

MATERIALS

- power supply, preferably >10 kV and >1 mA, current-limited
- two bars at least a meter long with an insulated mounting base

PROCEDURE

The Jacob's ladder[1] [1] is a device with little practical use but that often appears as a prop in old science fiction movies. It gets its name from the biblical character Jacob who had a dream of a ladder to heaven with angels ascending and descending on it.[2] It consists of two vertical conducting bars close together at the bottom (about ¼ inch) and far apart at the top (1 to 3 inches), like a pair of television rabbit-ear antennas. The bars need to be smooth to keep the arc moving. A high voltage applied between the bars causes an electric arc to form at the bottom where the bars are closest together. The arc then slowly rises up the bars until it finally extinguishes, whereupon it instantly strikes again at the bottom. Spirals (like a DNA molecule) and other decorative designs are also possible. The Jacob's ladder is sometimes called a "climbing arc."

To make a good Jacob's ladder requires a voltage in excess of 10 kV and a current in excess of about 10 milliamperes. The voltage primarily determines the length of the arc, and the current determines its intensity. The device will run with either AC or DC current. The voltage source must be such that it will withstand a short circuit indefinitely without harm. Transformers designed for neon signs are good because they have a large leakage inductance between their primary and secondary windings so as to limit the short-circuit current. With a normal transformer, you can place a suitable impedance

---

[1] Available from Carolina Biological Supply Company, Frey Scientific, Sargent-Welch, and Ward's Science.

[2] Genesis 28:12.

(inductance preferred, but resistance or capacitance will do) so as to limit the output current to the rated value. For example, if you have a transformer with an output rated at 20 kV and 10 mA and if the input is 115 V, the desired impedance is $115^2/(0.01\times20{,}000) = 66$ ohms in series with the primary winding. This impedance could be either a resistor with a power rating of at least $115^2/66 = 200$ watts, an inductor of $66/2\pi f = 175$ millihenries (for $f = 60$ Hz), or a capacitor of $1/(66\times2\pi f) = 40$ microfarads. The inductor must have enough iron so as not to saturate and a reasonable Q at $f = 60$ Hz (much less than 66 ohms of winding resistance). The capacitor must have a voltage rating greater than $115\sqrt{2}$ and not be electrolytic but must be designed for AC service. A bank of motor-starting capacitors connected in parallel is suitable. Another option is to place a large light bulb (100 watts or more) in series with the primary of the transformer.

You can introduce the demonstration by turning on the Jacob's ladder and asking where you have seen one of these. Someone will usually say "an old Frankenstein movie," and you can point out that they often appear as props in old science fiction movies, but that you will rarely see one in a modern physics laboratory. In fact, they have limited practical use except to dissipate the energy when an inductive circuit is interrupted either intentionally or by accident.

DISCUSSION

When you turn the device on, the full voltage initially appears across the bars, causing an arc to form at the closest point, where the electric field is highest. The voltage then drops to a low value because the impedance of the arc is low. The arc heats the air, and the hot air begins to rise because it is less dense than the surrounding air. As the arc rises, its impedance increases, and the voltage between the bars rises slightly. Eventually the arc gets so long that the voltage is not sufficient to sustain it, and the arc goes out. Then the voltage rises to its full open-circuit value, and the arc strikes again at the bottom. There is also a magnetic force causing the arc to rise, but the dominant effect is the heating of the air as you can demonstrate by tilting the whole device on its side.

The Jacob's ladder illustrates electrical discharges and the variation of the density of air with temperature. The principle is used in the horn gap on power transmission lines and in transformer yards for dissipating the arc of disconnects. It could serve as an introduction to a discussion of the physics of plasmas (ionized gases) and electrical breakdown. If you operate the Jacob's ladder inside a transparent closed insulated tube or box, the tube fills with a red-brown oxide of nitrogen. You can then aspirate the oxide through water to form nitric acid, illustrating the fixation of atmospheric nitrogen by lightning. The Jacob's ladder also converts oxygen gas (O_2) into ozone (O_3), a pungent, highly reactive allotrope of oxygen.

HAZARDS

This demonstration is potentially very dangerous. If currents of more than a few milliamperes are used, the Jacob's ladder will severely shock anyone who comes too close to the bars while it is in operation. Grounding the center tap of the transformer secondary provides some measure of safety, since it reduces the maximum voltage on the rods to half of what it would otherwise be. The safest arrangement would be with the whole apparatus enclosed in a transparent case, which can also serve to support the bars at the top, prevent air currents from extinguishing the arc, and allow you to fill the space with gases other than air to change the color of the arc. Be sure the insulating materials are not flammable. The case also acts as a resonant acoustical cavity, enhancing the sound of the arc, especially if it is designed to be resonant at the power-line frequency of 60 Hz (or 50 Hz in much of the world) or its harmonics. The bars can become very hot after prolonged operation. The Jacob's ladder produces radio-frequency emissions that can interfere with radio reception, and so it should not be operated continuously.

REFERENCE

1. H. Strand, *Popular Science* **184**, 110 (1964).

4.6 Tesla Coil

A Tesla coil, because of its high frequency, provides a relatively safe way to demonstrate phenomena associated with very high voltages and currents.

MATERIALS

- Tesla coil[1]
- fluorescent tube or clear light bulb
- insulated platform

PROCEDURE

The Tesla coil is a type of air-core resonant transformer capable of producing high-frequency voltages of upwards of a million volts. The Tesla coil was developed by Nikola Tesla (1856 – 1943), a contemporary and rival of Thomas Edison (1847 – 1931). Tesla's biography makes especially interesting reading [1–4]. In 1899 Tesla produced 135-foot-long discharges, 200 feet above the Earth with a 12-million-volt coil at his Colorado Springs laboratory, and the overload on the power line set fire to the alternator of the Colorado Springs Electric Company. Tesla imagined using his invention not only for wireless communications around the world but also for the widespread dissemination of electric power without the use of wires.

Because of its high frequency, the Tesla coil provides a relatively safe way to demonstrate very high voltages. A large Tesla coil is probably the most spectacular of all electrical demonstrations. The Tesla coil also makes an excellent student project since there is much room to experiment and to optimize its performance. The followers of Tesla almost constitute a cult. Old issues of electronic hobby magazines [5–10], scientific journals [11–18], and books [19–21] have construction details. A book that contains much practical design information is *Modern Resonance Transformer Design Theory*, by D. C. Cox.[2]

There are many other possible uses for the Tesla coil aside from producing long and impressive sparks in the air surrounding the top of the coil. You can illuminate a fluorescent or gas-filled tube or even a clear light bulb held in your hand some distance

---

[1] Available from Carolina Biological Supply Company, Fisher Science Education, Frey Scientific, Sargent-Welch, Science First, Ward's Science, and Scientifics.

[2] Available from Resonance Research Inc., E11870 Shady Lane Rd., Baraboo, WI 53913, 608-356-3647. This company also produces custom-made Tesla coils and other electrical demonstration apparatus.

away. A hydrogen-filled balloon on the end of an insulated rod will explode when you bring it near the top of the coil. A metal pinwheel attached to the high-voltage terminal of the coil will rotate as a result of the corona discharge. It is possible to draw the current through your body by bringing a long metal rod held in your hand near the top of the coil. You can use a chain of volunteers connected to the Tesla coil with a fluorescent lamp somewhere in the chain. With a large coil you should be able to light an incandescent bulb in series with the rod. You should use the highest voltage bulb possible to get the most light for a given current.

If a sufficiently well insulated platform is available, you should be able to stand with your bare feet on a metal plate (or more impressively sit in a washtub full of water) connected by a wire to the top of the coil and light a discharge tube or clear incandescent light bulb held in your hands or draw sparks off the tips of your fingers. In doing this demonstration, your hair should be wetted and metal thimbles worn on your fingers to prevent burns. A metal crown with points (lightning rods) worn on your head will prevent painful sparks from coming off the top of your head and reduce the remote possibility of catching your hair on fire. Sharply pointed metal is an advantage with the Tesla coil, unlike the Van de Graaff generator (see section 4.2), where it tends to discharge the subject. Use an experienced assistant for this demonstration rather than a volunteer. Give the subject a special name such as "The Human Lightning Rod" or "The Human Conductor" or "The Human Voltmeter." Caution people against trying electrical experiments like this at home.

The Tesla coil is essentially a radio transmitter without the antenna, and thus Tesla arguably deserves some credit for invention of the radio, although his interest was more in the transmission of electric power than in communication. It is capable of producing severe radio interference, and thus you should operate it inside a shielded cage or only for brief intervals. A suitable detector, such as a large loop of wire with a pair of conducting balls on the ends between which sparks can jump, placed some distance from the coil, demonstrates the propagation of these radio waves. You can also point out that the discharge is an example of a fractal (see section 6.13). A surprise visit by someone impersonating Nikola Tesla is a nice touch, and he can explain the operation of the coil and give a bit of history about the development of the Tesla coil and other of his inventions.

DISCUSSION

The Tesla coil consists of primary and secondary resonant circuits tuned to the same frequency

$$f = \frac{1}{2\pi\sqrt{LC}}$$

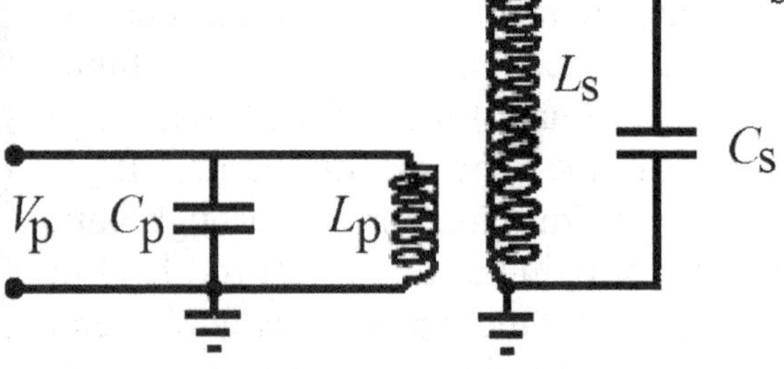

which is the order of 100 kHz to 1 MHz and designed to have the highest possible Q (lowest circuit loss). It is an electrical analog of the coupled pendulums (see section 1.18). The secondary capacitance C_s is a combination of the distributed capacitance due to the proximity of adjacent windings and the self-capacitance of the discharge electrode. The Tesla coil is normally designed to operate at the lowest resonant frequency of the secondary to avoid a voltage maximum other than at the top of the secondary coil. If the coupling k between the primary and secondary is adjusted to be near the critical value ($k \sim 1/Q$), then the secondary voltage is given approximately by

$$V_s = V_p\sqrt{C_p / C_s}$$

independent of the turns ratio [22]. In practice you would normally construct the secondary coil and then adjust C_p to give the largest V_s for a given V_p. You will have to retune the Tesla coil depending on what you connect to the secondary. Taps on the primary coil are useful for this purpose.

You can produce the input voltage V_p in a number of ways. One way is to connect the input through a rotary spark gap to the secondary of a high-voltage (several-kV) transformer connected to the AC power line through a current-limiting inductor. The opening and closing of the rotary spark gap (typically a few hundred pulses per second) creates a ringing voltage waveform after each opening. The spark gap and transformer can also be inserted in series between C_p and L_p to create a ringing waveform after each closing of the switch. You can replace the acoustically noisy spark gap with a thyratron or ignitron triggered from a source that is phase-locked to the 60-Hz (or 50-Hz in some parts of the world) power line. You can also derive the voltage V_p from a high-voltage, vacuum-tube oscillator circuit.[3] The voltage is frequently estimated from the length of the spark, but such a measurement is highly inaccurate due to the shape of the electrode and often leads to exaggerated claims of a million volts or more.

HAZARDS

People often claim that one can touch the secondary of the Tesla coil without being electrocuted because of the skin effect. High-frequency currents flow on the surface of the skin rather than through the body where damage to the heart could occur. However,

[3] See an old issue of *The ARRL Handbook for Radio Communications*, published yearly by the American Radio Relay League, 225 Main Street, Newington, CT 06111, 860-594-0200, http://www.arrl.org.

for the frequencies normally used with Tesla coils (less than 1 MHz), the skin depth typically exceeds the size of the body. Humans have a bulk resistivity of about 4 ohm-meters, giving a skin depth of about 2.3 meters at a typical frequency of 200 kHz. In fact, the muscles and nerves of the body are less sensitive to high frequencies than they are to DC or low-frequency AC. However, internal heating can occur and cause considerable pain, and the resulting burns heal very slowly.

Furthermore, it is probably best not to let anyone with a weak heart or wearing a heart pacemaker come into contact with the high voltage. In fact, with large Tesla coils, the use of volunteers is not recommended. In some Tesla-coil designs the primary has 60-Hz voltages present, and these could be lethal. A more common danger is the possibility of burns to the skin, especially if the discharge strikes the skin directly. For this reason, you should only draw the discharge to the body through a metal object that makes a large area of contact with the skin. For even more safety, you can wear a metal suit. If the discharge strikes an unsuspecting person, the result would surely be startling, and the recoil could cause injury.

If you use the Tesla coil to light a fluorescent tube, take care not to break the tube. The powder that coats the inner wall of such tubes and the mercury that they contain are highly toxic. In addition, the ozone and nitrous oxides produced by the discharge can corrode lung tissue even at levels too small to smell, and thus you should avoid prolonged exposure in a poorly ventilated room.

REFERENCES

1. J. J. O'Neill, *Prodigal Genius: The Life of Nikola Tesla*, Neville Spearman: London (1944).

2. I. Hunt and W. W. Draper, *Lightning in his Hand: The Life Story of Nikola Tesla*, Sage Books: Denver (1964).

3. T. C. Martin, *The Inventions, Researches and Writings of Nikola Tesla*, Omni Publications: Hawthorne, CA (1977).

4. M. Cheney, *Tesla: Man Out of Time*, Prentice-Hall: Englewood Cliffs, NJ (1981).

5. L. Reukema, *Experimenter* **4**, 309 (1925).

6. R. C. Dennison and E. H. McClelland, *Electrical Journal* **28**, 328 (1931).

7. K. Richardson, *Popular Electronics* **12**, 72 (May 1960).

8. C. Caringella, *Popular Electronics* **21**, 29 (Jul 1964).

9. E. Kaufman, *Popular Electronics* **21**, 33 (Jul 1964).

10. V. Vollono, *Popular Electronics* **71**, 29 (Aug 1989).

11. G. Breit and M. A. Tuve, *Nature* **121**, 535 (1928).

12. G. Breit, M. A. Tuve, and O. Dahl, *Phys. Rev.* **35**, 51 (1930).

13. C. R. J. Hoffmann, *Review of Scientific Instruments* **46**, 1 (1975).

14. I. Boscolo, G. Brautti, R. Coisson, M. Leo, and A. Lunches, *Review of Scientific Instruments* **46**, 1535 (1975).

15. D. G. Bruns, *Am. J. Phys.* **60**, 797 (1992).

16. K. D. Skeldon, A. I. Grant, and S. A. Scott, *Am. J. Phys.* **65**, 744 (1997).

17. K. D. Skeldon, A. I. Grant, G. MacLellan, and C. McArthur, *Eur. J. Phys.* **21**, 125 (2000).

18. M. Denicolai, *Review of Scientific Instruments* **73**, 3332 (2002).

19. H. S. Norris, *Induction Coils: How to Make, Use, and Repair Them*, Spon and Chamberlain: New York (1909).

20. G. F. Haller and E. T. Cunningham, *Tesla High Frequency Coil: Its Construction and Uses*, D. Van Nostrand Company: New York (1910).

21. A. D. Bulman, *Models for Experiments in Physics*, Crowell Company: New York (1966).

22. J. B. Earnshaw, *An Introduction to A-C Circuit Theory*, St. Martin's Press: New York (1960).

4.7 Faraday Cage

A screen cage large enough for a person to enter used with a Tesla coil or Van de Graaff generator illustrates the fact that a closed conducting surface is an equipotential.

MATERIALS

- closed screen cage
- Tesla coil or Van de Graaff generator
- two fluorescent tubes (optional)
- transistor radio (optional)
- cell phone (optional)
- metallic tinsel (optional)

PROCEDURE

The Faraday cage is simply a cage constructed of metal or screen that completely encloses its interior with an electrically conducting boundary. A Faraday cage large enough to enclose a person allows even more dramatic demonstrations. In such a case, provide a door to allow the person to enter and exit. In 1837, Michael Faraday performed an experiment in which he had someone enclose him with his field-measuring instruments in a large (12-foot) conducting cube charged to a high voltage until sparks flew from its corners, and proceeded to convince himself that he could not detect the fact that the cage was electrically charged. The Faraday cage is often used in conjunction with a Tesla coil (see section 4.6) or Van de Graaff generator (see section 4.2).

The most spectacular demonstration is to put a person inside a grounded Faraday cage and turn on a large Tesla coil or Van de Graaff generator outside the cage. It is especially good to use a celebrity known to the audience in the cage. For a school group, use a teacher. Make a comment about having a seat in what looks very much like an electric chair.

The discharge from the Tesla coil or the sparks from the Van de Graaff generator strike the cage but do not hurt the person inside. The person inside can hold a fluorescent tube that will not light and can even touch the inside of the cage without harm. The electric currents flow on the outside of the cage, and the interior of the cage is all at the same electric potential. The person should not stick a finger or other object through a hole in the cage, however. If you give someone in the front row a fluorescent tube, it will illuminate while the one in the cage does not. The demonstration is most effective in subdued illumination. After the demonstration, you can point out that the person in the cage was in fact safer than those in the front row.

With a sufficiently well-constructed Faraday cage, it should be possible to put a portable transistor radio inside and not be able to receive any radio stations on either the AM or FM bands. To do this requires electrically tight seams, holes in the conductor that are not too large, and the radio not too close to the walls of the cage. You could also call for a volunteer with a cell phone, and then ask them to try to make a phone call from within the cage. This method provides a sensitive way to test the quality of the Faraday cage.

With a sufficiently large Van de Graaff generator or Tesla coil and an insulated platform for the cage to rest on, it is possible to raise the voltage of an ungrounded cage to a high level with a person inside. Metallic tinsel such as used to decorate Christmas trees when attached to the outside of the cage will stick out like the hair of a person on a Van de Graaff generator, but similar tinsel inside the cage will show no response.

DISCUSSION

The Faraday cage illustrates a number of important safety considerations for someone caught in a lightning storm or required to work with high-voltage electricity. The safest place to be in a lightning storm is inside a closed metal container such as a building with a steel frame, an automobile (but not because of the rubber tires, which are far too small to prevent a lightning strike and contain carbon black, making them relatively poor insulators), or an airplane. A chicken coop or a garbage can with a lid would also do! One of the functions of a lightning rod is to shield the interior of a building by carrying the corona currents from the air around the outside of the building through a grounded conductor, thereby making the building all at the same electrical potential. High-voltage electrical circuits are enclosed within conducting boxes to protect those outside the enclosure. The metal enclosure forms an equipotential surface at the same voltage as the Earth and prevents any of the voltages inside the cage from reaching the outside.

Lightning kills an average of 73 people each year and injures over 300, far more than any other weather event, including tornadoes. A typical lightning bolt has a voltage of about 100 million volts and carries a current of 10–20 kA in a succession of several strokes, each with a duration of about 200 ms, separated by about 50 ms. Approximately 80% of lightning discharges remain within the cloud and never reach the Earth, but 25 million lightning strikes hit the ground each year worldwide.

Lightning also has some beneficial effects. It helped in the formation of amino acids, a precursor of life. It fixes nitrogen in the air, creating a natural fertilizer. It starts forest fires, which thin and renew the forest. There are stories about lightning restoring the sight to blind persons and even published claims of victims with improved intelligence.

The Faraday cage is named after Michael Faraday (1791 – 1867), a self-educated, English physicist and chemist whose lectures for the public in the 1840s became so popular that they helped save the Royal Institution of Great Britain from near bankruptcy

[1–5]. His lectures were attended by Charles Dickens (1812 – 1870) and by Prince Albert (1819 – 1861), the husband of Queen Victoria (1819 – 1901), and Prince Edward (1841 – 1910), her son (later Edward VII). Faraday so inspired Prince Albert that he studied chemistry at the University of Edinburgh. In 1826, Faraday also began a series of immensely popular, special Christmas lectures, originally for children, which have continued up to the present. Since 1966, these lectures have been televised and broadcast throughout Britain. When asked by Prime Minister William Gladstone (1809 – 1898) what was the use of electricity, Faraday replied, "Sir, I do not know, but some day you will tax it."

HAZARDS

There are no safety hazards to the person inside the cage as long as the cage is tightly sealed and no part of the body protrudes through holes in the cage. For the safety of those outside the cage, ground the cage thoroughly, but this is not required to protect the person inside. With a large Tesla coil, the experience can be frightening, and so you should only do it with a brave and willing volunteer.

REFERENCES

1. J. Kendall, *Michael Faraday: Man of Simplicity*, Faber & Faber: London (1955).
2. L. P. Williams, *Michael Faraday: A Biography*, Chapman and Hall: London (1965).
3. G. Caroe, *The Royal Institution: An Informal History*, John Murray: London (1985).
4. J. M. Thomas, *Michael Faraday and the Royal Institution: The Genius of Man and Place*, Adam Hilger: Bristol (1991).
5. B. L. Lan and J. B. S. Lim, *Phys. Teach.* **39**, 32 (2001).

4.8 Gas Discharge Tube

A partially evacuated glass tube filled with various gases at low pressure and connected to a high-voltage electrical source illustrates properties of electrical discharges and plasmas.

MATERIALS

- gas discharge tube[1]
- vacuum pump
- high-voltage power supply, current-limited
- various nonflammable gases
- pressure, voltage, and current meters (optional)
- large horseshoe magnet (optional)
- Geissler tubes[2] (optional)
- plasma globe (optional)
- spectrometer or diffraction grating (optional)
- Geiger counter (optional)

PROCEDURE

Electrical discharges in low-pressure gases make colorful and educational demonstrations [1–4]. The apparatus required consists of a glass tube about a meter long and at least several centimeters in diameter. Seal off the ends of the tube with rubber stoppers. One of the stoppers contains a hole through which you can insert a glass tube to allow you to fill the tube with various gases at reduced pressure. Pass electrodes such as nails through the stoppers, and connect them to a high-voltage, low-current source of electricity. Direct current is preferred, but you can also use 60 Hz (or 50 Hz in much of the world) from the power lines or even higher frequencies from a small Tesla coil. A voltage of 1,000 volts or more is required, but the current can as small as a few

---

[1] Available from American 3B Scientific and Klinger Educational Products.

[2] Available from Carolina Biological Supply Company, Frey Scientific, and Sargent-Welch.

milliamperes. Higher currents produce a brighter discharge but increase the danger of electrocution. The discharge is most stable if the power supply has relatively high output impedance, since the current drawn by the discharge is a sensitive function of the applied voltage. You can evacuate the tube with a water aspirator, but a mechanical vacuum pump capable of producing a vacuum of a few millitorr is preferred. Pressure, voltage, and current meters visible to the audience are useful additions.

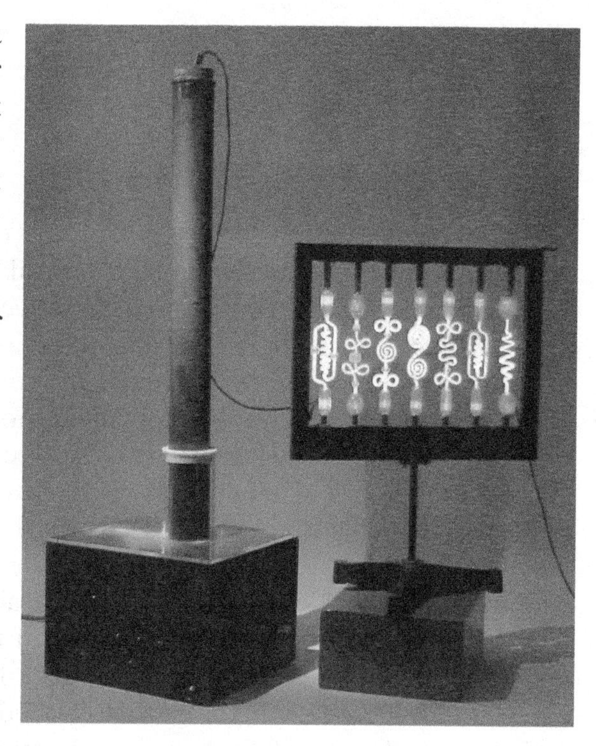

If a vacuum pump is not available, you can show some of the effects using a pre-sealed discharge tube such as a fluorescent lamp [5] or other special gaseous discharge lamps such as a Geissler tube, developed and popularized by the German instrument maker Heinrich Geissler (1814 – 1879), or a plasma globe (also known as a plasma sphere, lightning globe, Nebula Ball, or Sunder Ball) [6], available from many sources. Use a simple optical spectrometer or diffraction grating to show that the glowing gas emits certain discrete wavelengths that one can use to identify the gas. You can also use a Geiger counter to show that plasmas are not normally radioactive.

Fill the tube initially with air at atmospheric pressure, and apply the voltage. Normally, there will be no glow from the tube and no current drawn. As you slowly evacuate the tube, you will reach a pressure where current abruptly begins to flow and sparks emanate from the electrodes. As you further reduce the pressure, a steady violet glow will appear throughout the tube. At even lower pressure, alternating light and dark bands (called "Crookes dark spaces" after William Crookes, 1832 – 1919) can be seen along the length of the tube, until eventually the glow is completely extinguished. Note that the striations are always convex toward the cathode. You can then turn the pump off, and repeat the progression in reverse as you bleed air back into the tube. You can repeat the demonstration with other nonflammable gases such as neon, carbon dioxide, and argon. Each gas will produce a glow with a distinctive color characteristic of the gas.

Place a large horseshoe magnet near the discharge to affect its shape and to show that a magnetic field will affect the electrical discharge. Point out that a glow discharge does not obey Ohm's law, but rather it tends to maintain a constant voltage independent of the current at a given pressure.

DISCUSSION

This demonstration illustrates how the electrical conductivity of a gas varies with pressure. Normally, gases are good electrical insulators because the molecules of the gas are electrically neutral, consisting of negative electrons bound to equally positive nuclei. The few free charges that do exist, perhaps as a result of cosmic rays, are accelerated by the electric field along the length of the tube and collide with neutral gas molecules, losing their energy before they are able to ionize them. As you reduce the pressure, the mean free path for slowing down of the free charges increases, and the charges acquire enough energy to ionize other gas molecules, producing additional free electrons, which in turn ionize others, resulting in electrical breakdown. The alternating light and dark bands result from electrons emitted from the cathode that alternately accelerate and then stop as they acquire sufficient energy to be absorbed by the gas. At even lower pressure, the mean free path will exceed the length of the tube, the free charges will rarely collide with the gas molecules, and the discharge will extinguish. The emitted light is a result of excitation of the gas atoms from their ground state to excited states, with the subsequent emission of photons of light as they decay back to their ground state. Since atoms of different chemical elements have different energy levels, the color of the discharge depends on the gas used.

This demonstration can serve as an introduction to the subject of plasma physics [7–9]. Plasmas represent the fourth state of matter, along with solids, liquids, and gases. Plasmas represent the end result of raising the temperature of any substance, most of which transform successively through each of the four states. Most substances begin to take on the characteristics of a plasma at a temperature of a few thousand degrees Celsius and become fully ionized as the temperature approaches a million degrees. Plasmas differ from ordinary gases in that they are electrical conductors, a difference that is as remarkable as the differences among the other three states. Plasmas are relatively rare on Earth but constitute upwards of 99% of the known matter in the universe. Examples of plasmas in nature include bolts of lightning, the Aurora Borealis (the northern lights) [10], the ionosphere of the Earth, the solar wind, and the material of the Sun and stars. Laboratory plasmas are of great current interest because certain atoms such as isotopes of hydrogen at temperatures in excess of 100 million degrees will release energy by the process of nuclear fusion, a potentially clean and abundant source of energy [11–17]. Magnetic fields provide the most promising method for confining the plasma so that it does not cool by coming into contact with the walls of the vessel in which it is confined. Contrary to what you might expect, the plasma does not usually melt the walls of its container because its density (and hence its heat content) is too low, despite the high temperature.

HAZARDS

The main hazard is electrical shock from the voltage connected to the tube. There is also a possibility of implosion of the glass tube when you evacuate it and explosion if you pressurize it above atmospheric pressure (not recommended). It is possible to receive a mild burn from keeping your hand too long on the outside of a plasma globe. Make sure the belt of the vacuum pump has a guard, or place the pump out of reach.

REFERENCES

1. F. A. Maxfield and R. R. Benedict, *Theory of Gaseous Conduction and Electronics*, McGraw-Hill: New York (1941).
2. C. L. Strong, *Scientific American* **198**, 112 (Feb 1958).
3. B. Z. Shakhashiri, *Chemical Demonstrations*, The University of Wisconsin Press: Madison, WI, Vol 2 (1985).
4. H. F. Meiners, W. Eppenstein, R. A. Oliva, and T. Shannon, *Laboratory Physics*, John Wiley & Sons: New York (1987).
5. N. R. Guilbert, *Phys. Teach.* **34**, 20 (1996).
6. N. R. Guilbert, *Phys. Teach.* **37**, 11 (1999).
7. F. F. Chen, *Introduction to Plasma Physics*, Plenum Press: New York (1984).
8. Y. Eliezer and S. Eliezer, *The Fourth State of Matter: An Introduction to Plasma Physics*, Adam Hilger: Bristol (1989).
9. R. Hazeltine and F. Waelbroeck, *The Framework of Plasma Physics*, Perseus Books: Reading, MA (1998).
10. J. H. Clemmons and R. H. Evans, *Phys. Teach.* **33**, 34 (1995).
11. A. S. Bishop, *Project Sherwood*, Addison-Wesley Publishing Company: Reading, MA (1958).
12. T. K. Fowler and R. F. Post, *Scientific American* **215**, 21 (Dec 1966).
13. J. L. Bromberg, *Fusion: Science, Politics, and the Invention of a New Energy Source*, MIT Press: Cambridge, MA (1982).
14. R. Herman, *Fusion: The Search for Endless Energy*, Cambridge University Press: Cambridge (1990).
15. T. K. Fowler, *The Fusion Quest*, Johns Hopkins Press: Baltimore, MD (1997).
16. H. Wilhelmsson, *Fusion: A Voyage Through the Plasma Universe*, Institute of Physics: London (1999).
17. T. C. Simonsen, *The Industrial Physicist* **6**, 14 (2000).

4.9 Chaotic Circuits

Specially constructed electrical circuits produce chaotic output that can be seen and heard.

MATERIALS

- specially constructed circuits
- amplifier and loudspeaker (optional)
- oscilloscope (optional)
- computer with sound card (optional)

PROCEDURE

You can construct a variety of chaotic electrical circuits [1–10] to demonstrate the period-doubling route to chaos, bifurcations, and chaos. One of the simplest such circuit consists of two or more coupled relaxation oscillators using neon bulbs (such as a type NE-51) or tunnel diodes [11]. With eight such bulbs, the flashing could represent the white keys in one octave of a piano, and you can use that idea with a simple computer program to play a crude kind of chaotic music. The flashing appears quite random and unpredictable, and seemingly never repeats. Under some conditions the circuit will exhibit the opposite phenomenon of synchronization, whereupon a collection of coupled nonlinear oscillators of similar frequency begin flashing in synchronism in analogy with fireflies and other biological organisms [12].

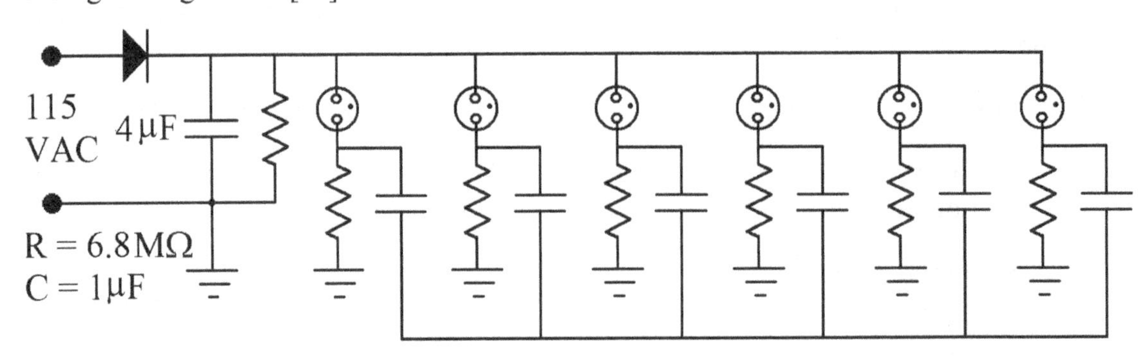

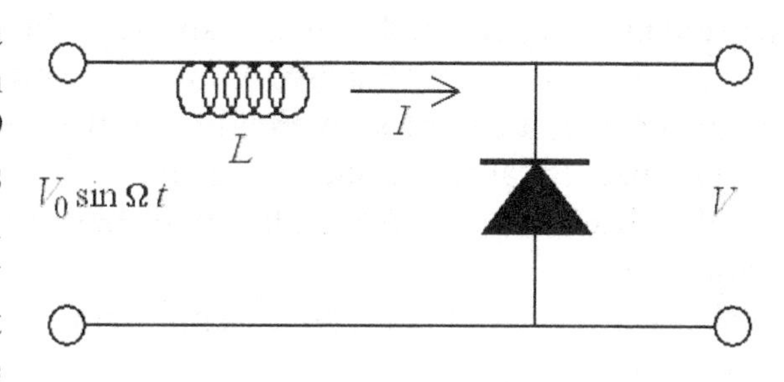

Another simple circuit consists of a silicon power diode with a large junction capacitance in series with a high-Q inductor of several hundred millihenries [13–18]. When driven with a sinusoidal voltage of at least 10 volts at a frequency in the range of 5–10 kHz, the output across the diode is chaotic with successive cycles of oscillation having different and apparently random amplitudes. You can display the output on an oscilloscope or connect it to an audio amplifier and loudspeaker to hear the period doubling and sound of chaos. You can also connect the output to an A-to-D converter or simply to the sound card of a computer [19] to collect data in digital form for more detailed analysis.

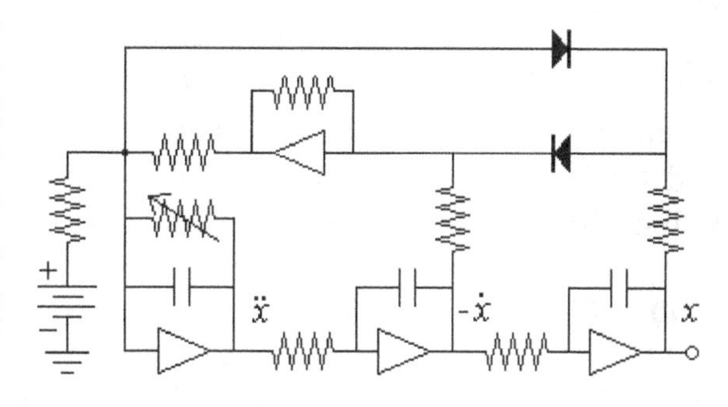

You can construct more complicated but more accurately controllable chaotic circuits with operational amplifiers using diodes or saturating amplifiers as the nonlinear element [20–22]. You can scale these circuits to any desired frequency and use them for careful analysis as part of a laboratory as well as for lecture demonstrations.

DISCUSSION

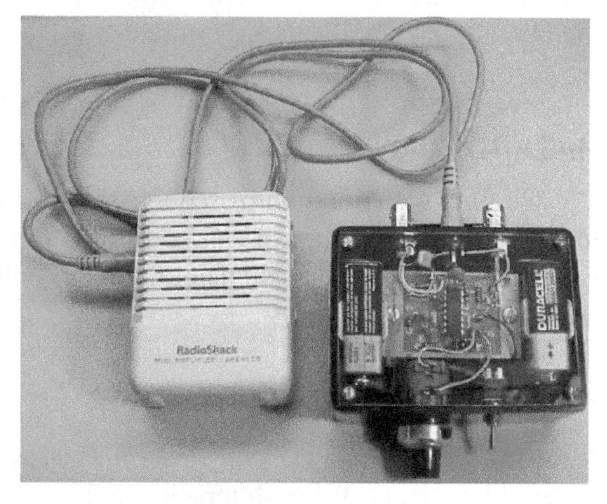

Chaos requires a nonlinearity. In the neon bulb circuit, the bulb is highly nonlinear. In fact, the bulb resistance is essentially infinite until the voltage across it exceeds a threshold of typically 80 volts, whereupon it drops to about 60 volts. The exact values depend on the type of gas, the nature of the electrodes, and the pressure. Thus the device is not only nonlinear, but it also exhibits negative resistance and hysteresis. A single such bulb in parallel with a capacitor charged though a resistor behaves as a relaxation oscillator, with the capacitor voltage rising slowly and then falling abruptly to a smaller voltage, whereupon the charging begins again.

The inductor-diode circuit works because the junction capacitance of the diode is a strong function of the voltage across the diode. Diodes have a junction capacitance when reverse-biased, but the main effect here is the time delay required for the charge to leave the junction when it first comes out of conduction. Diodes designed for high-current applications have more such charge and larger equivalent capacitance, sometimes reaching over a microfarad, which allows the circuit to work in the audio range. Low-capacitance signal diodes will also work in this circuit, but chaos will require drive frequencies of several hundred kilohertz, well out of the audible range. It is possible to simulate such a circuit numerically, but the equations are stiff and hard to solve without special numerical techniques. You can omit the audio oscillator by placing the circuit in the feedback loop of the audio amplifier and using the amplifier gain as the control parameter.

The circuit with operational amplifiers is an analog computer that solves the equation

$$\frac{d^3x}{dt^3}+A\frac{d^2x}{dt^2}+\frac{dx}{dt}-|x|+1=0$$

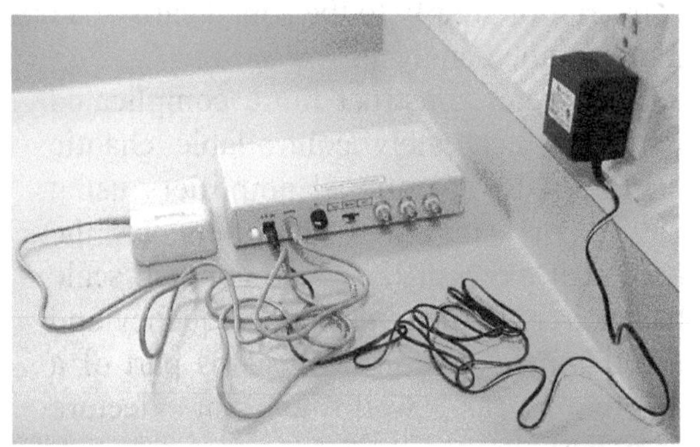

which has chaotic solutions for $A \cong 0.6$ [23]. The variable resistor in the circuit has a value $1/A$ and serves as the bifurcation parameter. If the variable resistor is a 2 kΩ potentiometer, the other resistors are 1 kΩ and the capacitors are 0.1 μF, the circuit will oscillate at a frequency of about 1,600 Hz, and the period-doubling bifurcations (800, 400, 200, ... Hz) will be easily audible. A low-pass filter in the output makes it even easier to hear the bifurcations. The battery voltage is arbitrary and only controls the amplitude of the output signal. Make it as large as possible without saturating the amplifiers, or use the positive power supply for the operational amplifiers with an appropriate series resistor. The power supplies and the grounded non-inverting inputs to the operational amplifiers are omitted from the schematic diagrams for simplicity.

HAZARDS

Circuits with neon bulbs require voltages of the order of 100 V, which is potentially lethal, although the current can be limited to safe values by a resistor. You can construct harmless, low-voltage versions of the neon-bulb circuits with tunnel diodes.

REFERENCES

1. F. C. Moon, *Chaotic and Fractal Dynamics: An Introduction for Applied Scientists and Engineers*, Wiley-Interscience: New York (1992).

2. M. A. van Wyk and W. -H. Steeb, *Chaos in Electronics*, Kluwer: Dordrecht (1997).

3. G. Chen and T. Ueta, Eds., *Chaos in Circuits and Systems*, World Scientific: Singapore (2002).

4. J. C. Sprott, *Chaos and Time-Series Analysis*, Oxford University Press: Oxford (2003).

5. T. Mishina, T. Kohmoto, and T. Hashi, *Am. J. Phys*. **53**, 332 (1985).

6. K. Briggs, *Am. J. Phys*. **55**, 1083 (1987).

7. T. Mitchell and P. B. Siegel, *Am. J. Phys*. **61**, 855 (1993).

8. M. T. Levinsen, *Am. J. Phys*. **61**, 155 (1993).

9. B. K. Jones and G. Trefan, *Am. J. Phys*. **69**, 464 (2001).

10. P. K. Roy and A. Basuray, *Am. J. Phys*. **71**, 34 (2003).

11. J. P. Gollub, T. O. Brunner, and B. G. Danly, *Science* **200**, 48 (1978).

12. S. H. Strogatz, *Synch*, Hyperion: New York (2003).

13. P. A. Linsay, *Phys. Rev. Lett.* **47**, 1349 (1981).

14. J. Testa, J. Perez, and C. Jeffries, *Phys. Rev. Lett*. **48**, 714 (1982).

15. R. W. Rollins and E. R. Hunt, *Phys. Rev. Lett.* **49**, 1295 (1982).

16. R. W. Rollins and E. R. Hunt, *Phys. Rev. A* **29**, 3327 (1984).

17. D. Smith, *Scientific American* **266**, 144 (Jan 1992).

18. R. V. Mancuso and E. M. Somerset, *Phys. Teach.* **35**, 31 (1997).

19. K. Hansen, M. Harnetiaux, and P. B. Siegel, *Phys. Teach.* **36**, 231 (1998).

20. J. C. Sprott, *Phys. Lett. A* **266**, 19 (2000).

21. J. C. Sprott, *Am. J. Phys*. **68**, 758 (2000).

22. K. Kiers, D. Schmidt, and J. C. Sprott, *Am. J. Phys.* **72**, 503 (2004).

23. S. J. Linz and J. C. Sprott, *Phys. Lett*. A **259**, 240 (1999).

5 Magnetism

Magnetism brings to mind horseshoe magnets and iron filings. However, magnetism is closely related to electricity. In 1819 the Danish physicist and chemist Hans Christian Oersted (1777 – 1851), during a lecture demonstration, observed that an electric current can affect a magnetic compass needle[1] and thus united what until then everyone had considered to be two distinct subjects and precipitated a flurry of activity [1]. The electric motor is the modern implementation of this phenomenon. In the 1820s, Michael Faraday[2] (1791 – 1867) in England and Joseph Henry[3] (1797 – 1878) in the United States independently demonstrated that a time-varying magnetic field can produce an electric current. Electric generators and eventually a world dominated by electronics and electrical devices was the result.

REFERENCE

1. J. Nelson, *Am. J. Phys.* 7, 10 (1939).

---

[1] The same discovery was previously made in 1802 by an Italian jurist, Gian Domenico Romagnosi (1761 – 1835), but it was overlooked because it was published in a newspaper, *Gazetta de Trentino*, rather than in a scholarly journal. Oersted was such a popular teacher and citizen that 20,000 people attended a torchlight procession at his funeral arranged by his students.

[2] Although nearly mathematically illiterate, Michael Faraday had excellent physical intuition and laboratory skills. He developed the concept of electric and magnetic fields and introduced a series of Christmas lectures for children at the Royal Institution in 1826 that still continues today.

[3] Joseph Henry was preceded only by Benjamin Franklin (1706 – 1790) as a highly acclaimed early American scientist. He was the first secretary of the Smithsonian Institution and made extraordinary contributions to the organization and to the development of American science.

Safety Considerations with Magnetism[1]

Although most public concern about electromagnetic fields has concentrated on power-line frequencies (50–60 Hz) [1, 2], radio frequencies (MHz) [3], and microwaves (GHz) [4], claims have been made that static magnetic fields can cause cancer or genetic damage. There is little theoretical reason to suspect that such fields might cause cancer or any other human health problems, and there is little laboratory or epidemiological evidence for a connection between static magnetic fields and human health (negative or positive) [5–8]. Nevertheless, several organizations have developed guidelines limiting continuous human exposure to 2,000 gauss and maximum exposure to 20,000 gauss, and to a lower level of five gauss for people with pacemakers or prosthetic devices. Humans routinely experience the Earth's magnetic field of a fraction of a gauss and commonly encounter fields of about two gauss on electric trains. A typical magnetic resonance imaging (MRI) scan subjects the patient to about 15,000 gauss.

The more serious danger is the large force than can exist between two magnets or between a powerful magnet and a ferromagnetic object, which can cause injury by pinching the skin or from shrapnel if the magnet shatters. Wear gloves and safety glasses when working with large rare-earth magnets.

REFERENCES

1. D. Hafemeister, Ed., Biological *Effects of Low-Frequency Electromagnetic Fields*, AAPT: College Park, MD (1998).

2. H. Takebe, T. Shinga, M. Kato, and E. Masada, *Biological Effects from Exposure to Power-Line Frequency Electromagnetic Fields*, IOS Press: Burke, VA (2000).

3. R. F. Cleveland and J. L. Ulcek, *Questions and Answers about Biological Effects and Potential Hazards of Radiofrequency Magnetic Fields*, Office of Engineering and Technology, Federal Communications Commission: Washington (1999).

4. J. R. Goldsmith, *Int. J. Occup. Environ. Health* **1**, 47 (1995).

5. J. E. Moulder and K. R. Foster, *Proc. Soc. Exp. Med. Biol.* **209**, 309 (1995).

6. J. E. Moulder, *IEEE Eng. Med. Biol.* **15**, 31 (Jul/Aug 1996).

7. K. R. Foster, L. S. Erdreich, and J. E. Moulder, *Proc. IEEE* **85**, 733 (1997).

8. J. E. Moulder, *Crit. Rev. Biomed. Engineering* **26**, 1 (1998).

---

[1] These results and precautions are excerpted mostly from the FAQ of Dr. John E. Moulder and the Medical College of Wisconsin at http://www.rareearth.org/magnets_health.htm.

5.1
Magnet and Cathode Ray Tube

A permanent magnet brought near a cathode ray tube causes a displacement or distortion of the pattern on the fluorescent screen to illustrate the effect of a magnetic field on moving charged particles.

MATERIALS

- permanent magnet
- oscilloscope or other operating cathode ray tube
- iron filings and overhead projector (optional)
- refrigerator magnet (optional)

PROCEDURE

You can illustrate the influence of a magnetic field on a moving charged particle by bringing a permanent magnet (bar or horseshoe) near the screen of a cathode ray tube (CRT). Preface this demonstration with a discussion of the poles of a magnet and the shape of the magnetic field near the magnet. The usual demonstration with iron filings, perhaps sprinkled on a sheet of transparent plastic above the magnet and projected on a screen with an overhead projector, helps the audience visualize the shape of the magnetic field. Point out that the magnetic field comes from the north pole of the magnet and loops around toward its south pole, though the direction is just a convention. You should also stress that the magnetic field lines do not terminate on the south pole of the magnet in the way electric field lines do on charges, but rather continue on through the iron and emerge at its north pole. Thus one cannot cut a magnet in half and isolate the north and south poles. It is not accurate to say that the field lines always close on themselves, but rather you should say that they cannot end. In practice, magnetic field lines often go to infinity or wander endlessly within some bounded region of space, although textbooks rarely show such examples. Repeat the demonstration with a thin refrigerator magnet in which the poles are close together [1].

Secondly, to make the demonstration effective, the audience must understand how a cathode ray tube works. You should stress the fact that the electron beam consists of a large number of negatively charged electrons moving at a good fraction of the speed of light (about $0.25c$ for a typical acceleration voltage of 17 kV). Make the analogy to the droplets of water exiting a garden hose. With a single spot on an oscilloscope screen, you can show the deflection of the spot when the magnet is brought near the screen. A horizontal line on the screen deflects upward at one end and downward at the other. The force points in the direction of the vector $q\mathbf{v} \times \mathbf{B}$, which, since the charge q of the electron is negative, is in the direction of $\mathbf{B} \times \mathbf{v}$, given by the right-hand rule. Alternately, working backward, you can determine which pole of a magnet is which by noting the direction in which the beam deflects.

With a pattern such as a sine wave on the oscilloscope, you can produce many interesting shapes with the magnet. The electron beam constitutes an electric current, with the current moving backward from the screen because of the negative charge of the electrons. In this way you can introduce the idea of magnetic fields exerting forces on currents and proceed to demonstrate the usual cases of forces on currents in wires. Another approach is to describe carefully the direction of the expected force using the right-hand rule, but to overlook the fact that electrons are negatively charged, and then ask the audience why the observed result is apparently backward. You could also discuss the fact that velocities are relative, and thus different observers would calculate different magnetic forces, unless the magnetic field itself is dependent on the reference frame, which in fact is the case [2].

DISCUSSION

There are many examples of the use of magnetic fields to deflect moving charged particles. Television and computer monitors (the old CRT type) use magnetic deflection rather than electric deflection as in most oscilloscopes. Charged particle accelerators such as cyclotrons use magnets to confine the particles in a circle while an electric field accelerates them, and plasma confinement devices such as the tokamak use complicated and intense magnetic fields to confine a gas of charged particles in the quest for controlled nuclear fusion. Other related examples are velocity analyzers, mass spectrometers, magnetrons, and Hall-effect devices. The role of magnetic fields in astrophysics (Van Allen belts, solar wind, cosmic rays, etc.) makes an interesting digression.

It may be tempting to try this demonstration using the picture tube of an old television set or the CRT monitor of an old computer. This is not advised because such tubes, especially color ones, require precise alignment of the electron beam and contain components that can become permanently magnetized and distort the picture. Old television sets and monitors contained a special circuit that degaussed the tube with an AC magnetic field that slowly decayed each time you turned on the set in an attempt to eliminate stray magnetic fields. Similar degaussers were used to bulk-erase audiotapes and computer disks.

HAZARDS

A potential hazard with this demonstration is breaking the glass on the CRT with a blow from the magnet. If the beam deflection is too small, it might be tempting to remove the CRT from its cabinet to get the magnet closer to the beam. You should not do this, since it is too easy to come into contact with the lethal high voltage used to accelerate the beam. The magnet can demagnetize credit cards, and can damage watches and other delicate instruments. It can attract to metal surfaces with sufficient force to pinch your finger and can adhere so tightly to such surfaces and to other magnets that you run a risk of injury trying to remove it.

REFERENCES

1. D. G. Haase, *Phys. Teach.* **34**, 60 (1996).

2. A. K. T. Assis and F. M. Peixoto, *Phys. Teach.* **30**, 480 (1992).

5.2 Eddy Currents

A permanent magnet dropped onto a copper plate bounces upward without touching the plate because of the eddy currents induced in the plate.

MATERIALS

- strong neodymium disk magnet[1]
- heavy piece of iron
- stainless steel disk of similar size and shape to the magnet (optional)
- copper plate, at least 2 cm thick
- liquid nitrogen, gloves or tongs, and safety glasses (optional)
- Styrofoam® container
- penny and overhead projector (optional)

PROCEDURE

There are many ways to demonstrate eddy currents [1, 2]. One of the easiest involves simply dropping a strong neodymium disk magnet onto a highly conducting copper plate. Show that it is a magnet by using it to lift a heavy piece of iron. Then drop it onto the plate of copper from a height of a few inches, letting the audience observe the bounce, followed by the magnet slowly settling onto the plate. Alternately set the magnet on its edge on the plate and tip it over, letting it slowly fall. Show that you can lift the magnet easily if you do so slowly, but that it is hard to lift it rapidly. Drop a stainless steel disk of similar size and shape to the magnet onto the plate to show that the magnetic field produces the effect.

If liquid nitrogen is available, cool the plate by placing it in a Styrofoam® container filled with liquid nitrogen to enhance the effect. Better yet, use a second copper plate that is already cold, since it will take a quarter hour or more for the plate to cool completely. You can observe the Leidenfrost effect (see section 2.10) when the liquid nitrogen is first poured over the plate. In a less expensive version of the demonstration, replace the

---

[1] Available from Arbor Scientific, Educational Innovations, Sargent-Welch, Scientifics, and Forcefield.

copper plate with aluminum, whose conductivity is slightly lower but increases with cooling in a way similar to copper.

While you have a strong magnet with a bath of liquid nitrogen, try to observe the paramagnetism that occurs when you suspend the magnet above the liquid. The effect presumably results from condensation of the oxygen in the air onto the droplets of liquid nitrogen, which by itself is not paramagnetic [3].

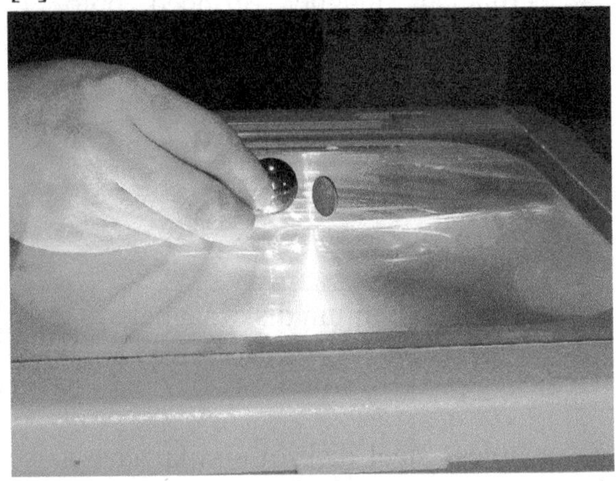

If a suitable copper or aluminum plate is not available, you can do a similar demonstration using a copper penny in place of the plate [4]. Balance the penny on its edge on an overhead projector and bring the magnet up to it. If you move the magnet rapidly, the penny will fall over, but if you move it slowly, the penny will remain upright, although some people may attribute the effect to air currents or vibration. After bringing the magnet slowly up to the penny, you can make the penny fall over by pulling the magnet away quickly. Pennies minted before 1981 work better because they are 95% copper, whereas more recent ones are copper-clad zinc and contain only 2.4% copper.

DISCUSSION

When the magnet approaches the copper plate, its magnetic field lines begin to penetrate the plate, inducing an opposing electrical current in the plate according to Lenz's law [5, 6]. This current produces a magnetic field that repels the other magnet. When you cool copper or aluminum to the temperature of liquid nitrogen (−196° C), its conductivity increases by about a factor of seven relative to its value at room temperature (20° C), and thus the eddy currents are stronger and decay more slowly. This experiment is possible because the neodymium-iron-boron ($Nd_2Fe_{14}B$) magnets developed in 1983 are capable of producing fields as high as 12,900 gauss. There is no theoretical reason why even stronger permanent magnets cannot be developed, and some day they probably will be.

HAZARDS

Rare-earth magnets break easily, especially if they snap together. They can snap together with sufficient force to pinch your fingers or skin. When they break, they can shatter and cut you. Strong magnets can destroy computer disks and credit cards, harm computer monitors, and potentially affect heart pacemakers. Liquid nitrogen is potentially dangerous since it can cause instant frostbite. Avoid letting it come into contact with bare skin. Use tongs or thermally insulated gloves to pick up the magnet or other objects that fall into the liquid nitrogen. Be careful that the liquid does not splash into your eyes. Wear gloves and safety glasses when working with liquid nitrogen. The copper plate will remain cold enough to cause severe injury and frostbite for nearly an hour after you remove it from the liquid nitrogen, unless you actively cool it with running water, for example, and even then, you may end up just warming the surface, and it will become cold again when you remove it from the water.

REFERENCES

1. T. D. Rossing and J. R. Hull, *Phys. Teach.* **29**, 552 (1991).
2. C. A. Sawicki, *Phys. Teach.* **34**, 38 (1996).
3. R. Simmonds, K. Browning, A. Rinker, T. Gastouniotis, and D. Ion, *Phys. Teach.* **32**, 374 (1994).
4. B. W. Holmes, *Phys. Teach.* **35**, 212 (1997).
5. W. M. Saslow, *Am. J. Phys.* **59**, 16 (1990).
6. W. M. Saslow, *Am. J. Phys.* **60**, 693 (1992).

5.3 Jumping Ring

A coil of wire wound around a short, cylindrical, laminated iron core is energized to propel a ring of aluminum up to the ceiling.

MATERIALS

- iron-core solenoid[1]
- push button with rugged contacts
- continuous ring of aluminum to fit over the iron core
- permanent magnet (optional)
- identical ring with gap in it (optional)
- iron rod (optional)
- dish of liquid nitrogen and safety glasses (optional)
- capacitor to resonate the solenoid at 60 (or 50) Hz (optional)
- light bulb and coil (optional)

PROCEDURE

The coil of wire (sometimes called "Thomson's coil" after Elihu Thomson, 1853 – 1937) can be operated directly from the 60-Hz (or 50-Hz) power line through a push button switch with contacts rugged enough to stand the arcing that occurs when the switch is opened. A 10-cm-diameter core about 20 cm long with several hundred turns of #16 wire provides sufficient magnetic field and inductance to propel the ring while limiting the current to a tolerable value. Point out that aluminum is not normally a magnetic material, and show that it is not attracted to a permanent magnet, but that the induced current flowing in it momentarily magnetizes it. You can make an aluminum ring by sawing off the end of an aluminum pipe [1].

[1] Available from PASCO Scientific.

You can place an iron rod or pipe, perhaps 40 cm long, on top of the core to concentrate and extend the magnetic field, causing the ring to jump much higher into the air. Make a lengthwise saw cut in the iron pipe to remove eddy currents. By cooling the ring in a dish of liquid nitrogen (−196°C), you can make the ring go even higher. If the ring has a gap in it, it will not move when you energize the coil. You can demonstrate continuous rings of different materials. If you do the demonstrations in the right order, you can induce the audience to plead at each step to make the ring go higher.

In an alternate version of the demonstration, place a capacitor of such a value as to make the circuit resonate at near 60 Hz (or 50 Hz in much of the world) in series with the coil. With careful tuning, you can make the ring oscillate up and down on the iron core by the variation in the inductance and corresponding current caused by the ring's position [2].

You can use the same apparatus to illustrate the principle of the transformer by connecting a small light bulb to a coil that you lower over the coil that propels the ring. If you choose the coil and bulb appropriately, it will light without burning out, and you can show that the iron core concentrates and extends the magnetic flux, showing why transformers have iron cores to improve the coupling between their primary and secondary windings.

DISCUSSION

The ring jumps into the air because of the current induced in it, which is in a direction opposite to the direction of the current in the coil [3−16]. The opposing currents repel one another. The magnitude of the induced current depends on the resistance of the ring. Thus a good electrical conductor such as aluminum or copper is required. Aluminum is preferred because of its smaller mass, which enables it to accelerate more easily. The resistivity of aluminum and copper is about a factor of seven lower when they are at the temperature of liquid nitrogen. The repulsion occurs because the conducting ring tends to exclude the magnetic flux that the coil attempts to force through it (Lenz's law). The ring does not attract to a permanent magnet, but rather the force results from the current induced in the ring by the changing magnetic field of the coil. If the ring were purely resistive, the current flowing in it would be 90° out of phase with the magnetic field, and there would be no net force on the ring. Because the ring has inductive reactance, which dominates the resistance, especially when the ring is cold, the current is closer to 180° out of phase with the magnetic field, producing a net force. If the ring has a gap in it, there will be no force on it because there is no path for the current to flow.

HAZARDS

The voltages are potentially lethal, and there is a dangerous high-voltage transient when the switch opens. The ring jumps with considerable force, and so you should keep the area above it clear. Do not do the demonstration beneath overhead lights that could shatter. Since liquid nitrogen can cause frostbite, handle the ring with tongs during that demonstration, and wear safety glasses. With practice, you can usually catch the ring with the tongs, to the audience's delight.

REFERENCES

1. E. R. Laithwaite, *Propulsion without Wheels*, Hart: New York (1968).
2. L. Strong, *Scientific American* **205**, 143 (Aug 1961).
3. R. M. Sutton, Ed., *Demonstration Experiments in Physics*, McGraw-Hill: New York (1938).
4. H. E. White and H. Weltin, *Am. J. Phys*. **31**, 925 (1963).
5. E. J. Churchill and J. D. Noble, *Am. J. Phys*. **39**, 285 (1971).
6. D. J. Sumner and A. K. Thakkar, *Phys. Educ.* **7**, 238 (1972).
7. G. D. Freier and F. J. Anderson, *A Demonstration Handbook*, AAPT: College Park, MD (1972).
8. W. R. Towler and J. W. Beams, *Am. J. Phys*. **44**, 478 (1976).
9. A. R. Quinton, *Phys. Teach.* **17**, 40 (1979).
10. S. Y. Mak and K. Young, *Am. J. Phys*. **54**, 808 (1986).
11. W. M. Saslow, *Am. J. Phys*. **55**, 986 (1987).
12. N. Thompson, *Thinking Like a Physicist*, Adam Hilger: New York (1990).
13. J. C. West and B. V. Jayawat, *IEEE Proceedings* **33**, 292 (1990).
14. T. D. Rossing and J. R. Hull, *Phys. Teach.* **29**, 552 (1991).
15. R. V. Mancuso, *Phys. Teach.* **30**, 196 (1992).
16. J. Hall, *Phys. Teach.* **35**, 80 (1997).

5.4
Can Crusher

A large capacitor discharged into a low-impedance coil of a few turns produces a magnetic field of strength sufficient to crush an aluminum soft drink can.

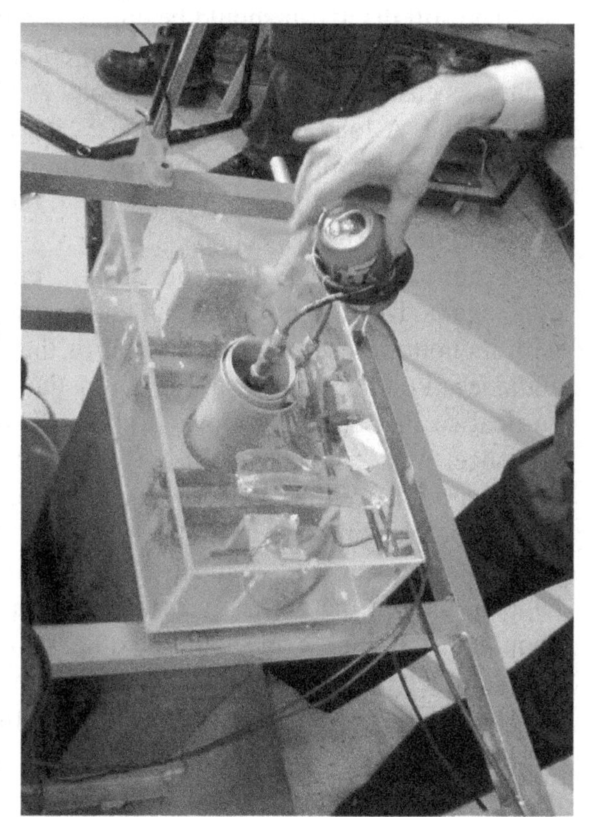

MATERIALS

- 60-μF, 10-kV capacitor with charging circuit[1]
- class-A ignitron, spark-gap, or mechanical switch
- coil of 5 turns of 1/4-inch-diameter copper tubing
- aluminum soft drink cans
- safety glasses
- large voltmeter, 10-kV full scale (optional)
- baseball glove (optional)
- liquid nitrogen (optional)

PROCEDURE

You can make an effective demonstration [1–3] of the strength of the magnetic force using a 60-μF, 10-kV capacitor and a coil consisting of 5 turns of quarter-inch-diameter copper tubing. The peak current can reach the order of 100,000 amperes and thus requires a special switch such as a class-A ignitron or spark gap. For safety, only charge the capacitor to 8 kV, and in fact the power supply should be designed so that it is incapable of charging the capacitor beyond its rated voltage for safety reasons. Wind the coil so that it fits tightly around a soft drink can with only a thin layer of insulation. The tight fit is necessary to induce the required currents in the can. The insulation must hold off 10 kV or more. The turn-to-turn voltage of the coil can exceed 1 kV, and thus you must insulate the turns from one another. The operation improves if you flatten the tubing in a vise to

[1] The same apparatus can be used in an exploding wire demonstration (see section 4.4).

reduce the height of the coil. In the simplest arrangement, push a button to initiate the charge, and the ignitron or spark gap triggers when the voltage reaches the desired level and you release the button. Alternately, use separate charge and discharge buttons. A voltmeter connected to the capacitor and clearly visible to the audience is a useful addition.

The soft drink can should be empty and have a hole in the top for the air to escape. In fact, it crushes more completely if you cut away the bottom of the can to allow the air to escape as easily and quickly as possible. A nice touch is to use a soft drink with "Crush" in its name. A loud noise accompanies the crushing of the can, and so you should warn the audience. If you quickly remove the crushed can, it will feel warm because of the currents induced in the can. You can hold the can up for the audience to see and then pass it around for closer inspection.

After crushing the can, you can ask the audience whether they would like to see it done again. This usually elicits an enthusiastic response. This time, however, rest the can with its bottom just touching the end of the coil, so that the can is propelled out into the audience. For this demonstration, aim the coil axis at a 45° angle with respect to the vertical so that the can follows a parabolic trajectory with maximum range out into the audience. A nice touch is to use a 7-Up® can with the bottom intact for this part of the demonstration. With practice, you can cause the can always to land in the same seat and ask the person sitting there to try to catch it. Handing the person a baseball glove before propelling the can adds a touch of drama and clues the audience in to what is about to happen without the necessity of an explanation. In practice, the person catches the can about half the time. You should be quick to apologize for aiming badly if the person fails to catch the can so that the audience does not embarrass the volunteer. "Take me out to the Ball Game" music is a nice touch while the capacitor is charging.

DISCUSSION

The magnetic field produced by the coil is proportional to the number of ampere-turns in the coil which is equal to $NV\sqrt{C/L}$, where N is the number of turns, V is the capacitor voltage, C is the capacitance, and L is the inductance of the coil and its leads back to the capacitor. The magnetic force is proportional to the square of the field or N^2V^2C/L. Thus an effective demonstration requires a large voltage and capacitance and a low inductance. It appears that increasing the number of turns is advantageous, but beyond a certain point the improvement is offset by the increase in inductance with N. Alternately, you can consider that when properly designed, essentially all the energy in the capacitor ($CV^2/2$) is transferred to the magnetic field whose energy is approximately $B^2/2\mu_o$ times the volume of the interior of the coil.

This demonstration illustrates a number of important physical principles. The can is crushed or propelled by the repulsive force between the current in the coil and the induced current in the can. Thus it illustrates both the force exerted on a current by a magnetic field and Faraday's law of induction. The magnetic pressure ($B^2/2\mu_o$) exists only in the space between the turns of the coil and the can in the brief time before the magnetic field soaks through the can. The electrical circuit consists of a damped harmonic oscillator. The can does not attract to a magnet, and the effect would not occur with a tin can because of its much higher electrical resistance. Cooling the can with liquid nitrogen decreases its resistance but does not provide much improvement, presumably because the can already crushes faster than the current dies away for a room-temperature, aluminum can. The necessity of letting the air escape illustrates the ideal gas law, the inverse relation of pressure and volume at constant temperature (Boyle's law), and the viscosity of the air that precludes its rapid expulsion. You could repeat the demonstration with cans having different size holes for the air to escape.

The crushing of the can is analogous to the plasma theta pinch. The ripples that occur around the circumference of the can when it crushes illustrate an instability well known with plasmas, and you can study the variation of the wavelength of the ripples with different types of cans. Very strong magnetic fields ($>10^7$ gauss) can be produced by the clever use of explosives and used to study matter under pressure conditions found in stars and planets [4]. The record for continuous field magnets is 3.31×10^5 gauss (33.1 Tesla) [5].

HAZARDS

Considerable energy is stored in the capacitor, and the voltages and currents used in this demonstration are potentially lethal. It is best to place the capacitor behind the lecture bench with only the coil exposed to protect the audience in the event of an explosion. You should not charge the capacitor above about 80% of its rated voltage. All high-voltage conductors should be well insulated. Stand well back while it is being charged and discharged. The capacitor should have a bleeder resistor or, better yet, a shorting bar that automatically shorts the capacitor through a low resistance (less than 100 ohms) whenever it is not being charged or waiting to discharge into the coil. Store the capacitor with its terminals shorted. The controls used to initiate the charge and discharge should be on a long cord and provided with a secure safety ground. A sufficiently powerful discharge can separate the can into two pieces, which are ejected from the coil with considerable force. The propelled can is relatively harmless unless you aim it directly at someone. Remove any sharp points and loose tabs from the can, however. Wear safety glasses for this demonstration as well as protective gloves if you cool the can with liquid nitrogen.

REFERENCES

1. R. J. Allen, *Am. J. Phys.* **30**, 867 (1962).
2. J. C. Thompson, *Am. J. Phys.* **31**, 397 (1963).
3. A. W. Desilva, *Am. J. Phys.* **62**, 41 (1994).
4. F. Bitter, *Scientific American* **213**, 65 (Jul 1965).
5. B. Keoun, *Science* **273**, 869 (1996).

5.5 Levitated Ball

Alternating current in a pair of magnet coils produces a magnetic field of a shape and strength that is sufficient to stably levitate an aluminum ball.[1]

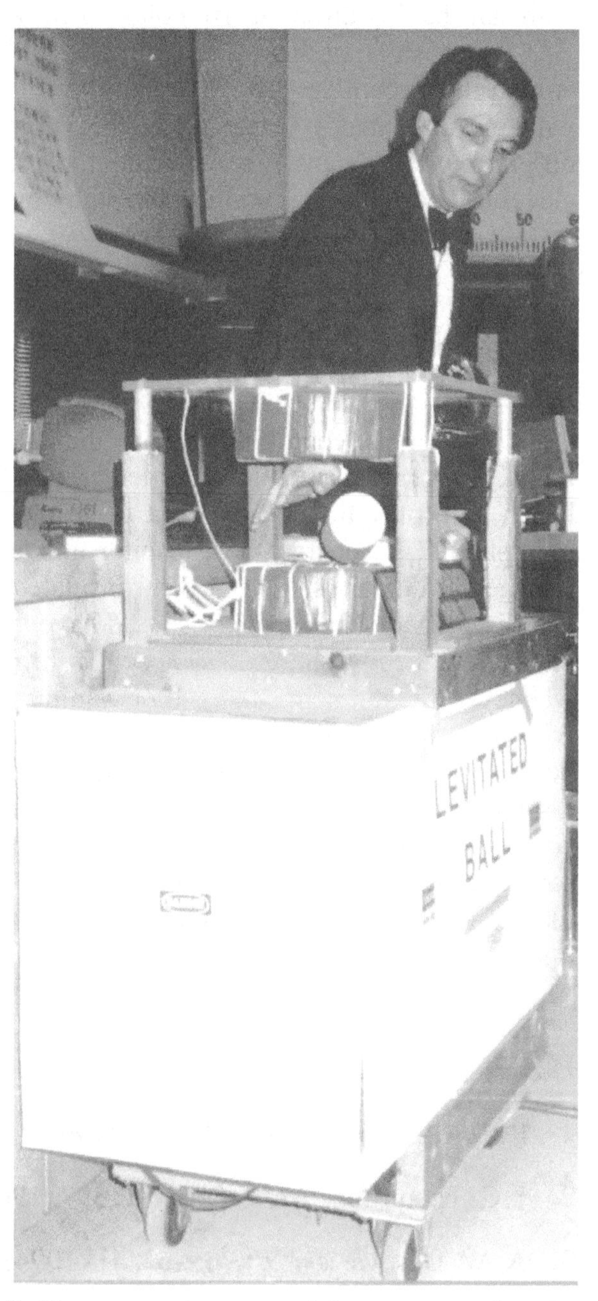

MATERIALS

- two magnet coils
- 60-Hz (or 50-Hz) power supply
- hollow aluminum ball or cylinder
- permanent magnet (optional)
- liquid nitrogen, gloves, and safety glasses (optional)

PROCEDURE

You can levitate an aluminum ball or other light, highly conducting object using the AC magnetic field of a pair of identical coils with their axes along a vertical line and separated by a distance comparable to their diameter. When the currents in the two coils flow in opposite directions, they produce a cusp-shaped field in which the magnetic field is zero at the center point between the coils and increases in every direction outward. This magnetic well is capable of stably supporting a 10-centimeter-diameter, hollow, 1-cm-thick, aluminum ball by virtue of the eddy currents induced in the ball. You can point out and demonstrate that the ball does not attract to a permanent magnet. A hollow, 1-cm-thick, aluminum cylinder 38 centimeters long by 8 centimeters in diameter will also levitate but at a lower equilibrium position. The cylinder executes remarkable gyrations and oscillations if set into motion. If

[1] Contributed by Professor Donald W. Kerst (1911 – 1993).

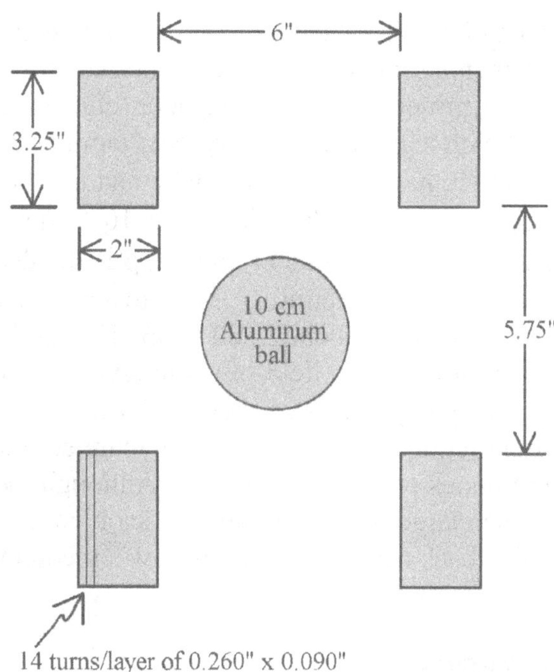

you cool the cylinder to liquid nitrogen temperature (−196°C) before placing it in the field, its resistivity drops by a factor of seven, and it will begin spinning rapidly for reasons that are not fully understood.

If you connect the coils so that their currents flow in the same direction, they form a so-called magnetic mirror, and a diamagnetic object like an aluminum ball has a stable position along the axis of the coils but not at right angles to the axis. Thus the ball is ejected sideways. Because of this behavior, it is good to hold the ball with a net bag made of a few strands of twine. The reason the ball is ejected is that the magnetic field decreases radially outward from the axis, and thus the ball rests in a magnetic well for motion along the axis, but it rests unstably at the top of a magnetic hill for motion perpendicular to the axis. Such an equilibrium condition constitutes a saddle point because it resembles the situation of a marble perched on the saddle of a horse. Potato chips sometimes have the same shape.

Many people confuse magnetism with gravity. Ask the audience if they think one would drift off into outer space if the magnetic field of the Earth disappeared. A surprising number of people think this is so, but of course the two forces have little to do with one another. You might point out that the Moon has almost no magnetic field, and that the magnetic field of the Earth is far too weak to attract a person, even if made of iron, although frogs have been magnetically levitated.

DISCUSSION

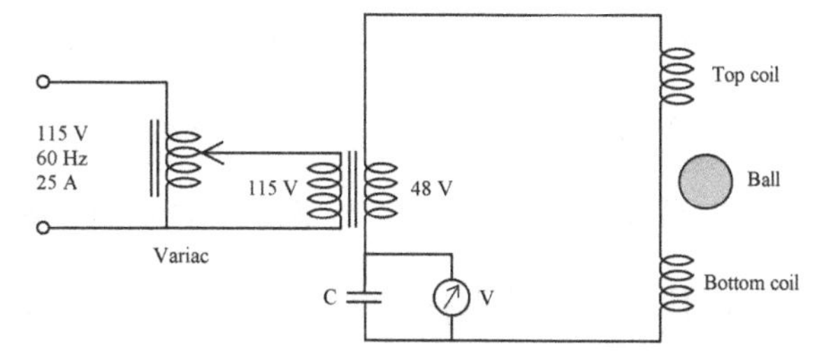

The construction of a magnetic levitation apparatus is difficult because large magnetic fields are required over a relatively small volume, and thus the coils must have a large current density and corresponding heat dissipation. For coils with a depth of about 5 centimeters in air, the cooling by free convection is sufficient for current densities of about 200 amperes per square centimeter. These coils run at a higher current density, and dimensions as shown will allow short (30-second) periods of operation. The capacitors resonate with the coil inductance at the power-line frequency (usually 50 or 60 Hz) so that the current drawn from the power line is much less than the circulating current in the coils. The power line requires a 25-ampere capacity. The capacitance required for the mirror connection (C = 170 μF) is slightly less than for the cusp connection (C = 212 μF) because of the mutual coupling between the coils. A Variac allows the ball to levitate when the voltmeter reads about 600 volts. The coil layers should be insulated, varnished, and baked. Electric motor repair shops can often fabricate such coils.

In principle, any diamagnetic object can levitate in a sufficiently strong magnetic field with a gradient [1]. The most familiar example is magnetically levitated trains. Most objects in nature are diamagnetic and can be levitated if the magnetic field gradient ∇B^2 exceeds $2\mu_0 g\rho/\chi$, where $\mu_0 = 4\pi \times 10^{-7}$ F/m is the permeability of free space, $g = 9.8$ m/s^2 is the acceleration due to gravity, ρ is the density of the material to be levitated, and χ is its magnetic susceptibility. Most materials have $\rho/\chi \cong 10^8$ kg/m^3, and hence the required field gradient is about 2,500 T^2/m. Thus to levitate an object with a size of 10 cm would require a magnetic field of about sixteen Tesla. Many objects have been levitated in this way including pieces of cheese, pizza, frogs, and even a mouse [2, 3]. Levitating a human would require a special racetrack magnet of almost forty Tesla and about one gigawatt of continuous power consumption. Although the magnetic force on ferromagnetic materials is much larger than on diamagnetic objects, they cannot stably levitate without some form of feedback according to Earnshaw's theorem [4].

HAZARDS

Insulate the coils and their electrical connections to prevent electrical shock. The coils should not be overheated. Keep watches and other items sensitive to strong magnetic fields out of the vicinity of the coils. Avoid dropping the aluminum ball, since it is rather heavy. Do not touch the ball when it is at the temperature of liquid nitrogen temperature. Wear safety glasses and gloves when working with liquid nitrogen. When the ball begins to spin rapidly, it can be ejected from the region of the magnetic field with considerable force. There are no known biological hazards from strong magnetic fields. Volunteers have spent up to forty hours inside a 4-Tesla whole-body magnet without any obvious ill effects [5].

REFERENCES

1. T. D. Rossing and J. R. Hull, *Phys. Teach.* **29**, 552 (1991).

2. M. V. Berry and A. K. Geim, *Eur. J. Phys.* **18**, 307 (1997).

3. A. Geim, *Physics Today* **51**, 36 (Sep.1998).

4. T. B. Jones, *J.App.Phys.* **50**, 5057 (1979).

5. J. F. Schenck, *Annals NY Acad. Sci.* **649**, 285 (1992).

5.6 Superconductors

High-temperature superconductors used with permanent magnets illustrate the Meissner effect.

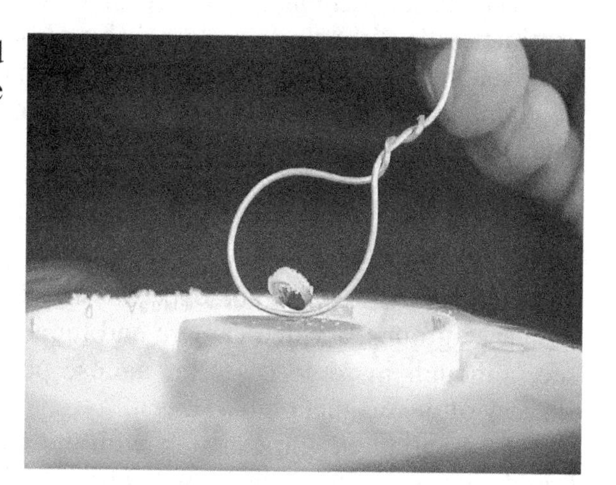

MATERIALS

- high-temperature superconductor[1]
- glass dish
- small, strong, permanent magnet
- small plastic tweezers
- piece of paper or small wire loop
- Dewar flask of liquid nitrogen
- gloves and safety glasses
- video camera and monitor (optional)
- pieces of plastic, iron, and copper (optional)
- bar magnet (optional)
- small compass (optional)

PROCEDURE

What for many years was an exceedingly difficult demonstration requiring liquid helium became rather ordinary [1] with the discovery in 1986 of materials such as yttrium-barium-copper-oxide ($YBa_2Cu_3O_7$) that become superconducting at temperatures above the boiling point of liquid nitrogen (77K) [2]. The results were presented at the March 1987 meeting of the American Physical Society in what the *New York Times* called "The Woodstock of Physics." These superconductivity kits are now available at low cost from a number of sources. You need only supply liquid nitrogen, which is available from hospital supply houses and other sources. At least one vendor[2] markets a

---

[1] Available from Carolina Biological Supply Company, Klinger Educational Products, Sargent-Welch, Ward's Science, and TEL-Atomic.
[2] Available from Arbor Scientific.

kit for the fabrication of superconductors requiring a kiln capable of reaching 1,000°C and a medium-sized vise. An alloy of mercury-barium-calcium-copper-oxide ($HgBa_2Ca_2Cu_3O_8$) has been found that becomes superconducting at temperatures of 135K at atmospheric pressure and at 164K at a pressure of 31 GPa [3].

The usual demonstration consists of placing a disk of the material several centimeters in diameter in a glass dish and putting just enough liquid nitrogen in the dish to cover the disk [4]. Then place a small magnet made of a rare-earth material such as neodymium[3] (NdFeB) or samarium-cobalt above the disk with tweezers and levitate it a few millimeters above the disk. Alternately, you can suspend the superconductor over a strong permanent magnet. It takes about a minute for the superconductor to cool sufficiently for the effect to occur. You can pass a piece of paper or a wire hoop under the magnet to illustrate that it is really levitated. Point out that magicians often do this with a levitated person. Ask the audience why magicians always use a woman rather than a man. For large groups, a video camera and monitor are almost essential to allow everyone to see, although in some cases an intense point source of light can suffice to project a shadow of the magnet onto a screen.

As a variation, you can start with the magnet resting on the disk before adding the liquid nitrogen. The magnet will abruptly levitate when the disk reaches the critical temperature. If you allow the nitrogen to boil away, the magnet suddenly falls, and you can repeat the whole process. If water vapor condenses from the air, it can form a layer of ice causing the magnet to stick to the disk.

When you remove the magnet from the vicinity of the superconductor, a residual current will remain in the superconductor as you can demonstrate by holding a small compass above the disk [5]. You can demonstrate that the effect persists as long as the superconductor remains sufficiently cold, illustrating the lack of electrical resistance that would cause the current to decay. As the superconductor warms up, the current will suddenly die, and the compass will reorient itself.

Another demonstration that is more visible to a large audience requires a disk of superconductor suspended by a string about half a meter long. Beside the superconductor, suspend on separate strings materials of similar size and shape but made of plastic, iron, and copper. First bring a strong bar magnet near the plastic with no effect. Then bring the magnet near the iron (try a steel washer), which is strongly attracted to the magnet. Point out that plastic is an electrical insulator and iron is an electrical conductor but not as good a conductor as copper. Then ask the audience to predict the effect when you bring the magnet near the copper. If you bring the magnet slowly up to the copper, there is no effect, but if you wave the magnet back and forth next to the copper, the copper disk begins to swing. You can then explain eddy currents generated by time-varying magnetic fields (see section 5.2).

Finally bring the magnet up to the warm superconductor. By waving the magnet back and forth, show the audience that there is no effect and that the material behaves like an electrical insulator. Then lower the superconductor into a Dewar flask of liquid nitrogen for about half a minute until the boiling stops, and repeat the demonstration. This time there is an obvious repulsion of the superconductor from the magnet. The effect persists for about a minute until the superconductor warms up again.

---

[3] Available from Arbor Scientific, Educational Innovations, Forcefield, and Sargent-Welch.

Another dramatic demonstration, suitable for slightly more sophisticated audiences, is to place a toroid of superconducting material around an iron transformer core [6]. Connect the transformer secondary to an incandescent lamp that ceases glowing when you cool the superconductor. The transformer primary needs to have enough series impedance to tolerate the shorted turn represented by the superconductor.

You should probably not conclude the demonstration without a comment about the potential importance of high-temperature superconductors and the difficulties of using the technology in practical applications [7]. Aside from the low temperatures required, the high-temperature superconductors are a form of brittle ceramic, and it is difficult to fabricate wires that are thin, flexible, and free of defects. Approximately 7% of the electrical power generated in the United States is lost in transmission line resistance, and superconductors could eliminate half of that waste [8]. Superconducting magnetically levitated trains in Japan reach speeds of 343 miles per hour. The military uses superconductors in applications including propulsion systems for ships, ultrasensitive detectors of submarines and underwater mines, and electromagnetic pulse generators for destroying power grids and electronics.

DISCUSSION

A magnet will levitate above a superconductor (or a superconductor above a magnet) because of the Meissner effect, more properly called the Meissner-Ochsenfeld effect after Walter Meissner (1882 – 1974) and Robert Ochsenfeld (1901 – 1993), who discovered it in 1933, and which expels the magnetic flux from a superconductor. The phenomenon of superconductivity was first discovered in mercury by Heike Kamerlingh Onnes (1853 – 1926) in 1911, for which he received the 1913 Nobel Prize in physics. Superconductivity was explained by John Bardeen, Leon Cooper, and Robert Schrieffer in what is commonly called the BCS theory, for which they received the 1972 Nobel Prize in physics.

The image one should have is of a magnet with its field lines emerging from one pole and looping around the outside to re-enter at the pole on the other end. Since these field lines cannot penetrate the superconductor, the magnet rises above the superconductor to give the field lines space to return. In reality, the superconductor traps some magnetic flux, and the resulting persistent current accounts for the residual magnetism. A superconductor is an example of a diamagnetic material, in contrast to iron, which is ferromagnetic. Diamagnetic objects attract to regions of weak magnetic field, whereas ferromagnetic (and paramagnetic) objects attract to regions of strong magnetic field.

HAZARDS

Do not allow liquid nitrogen or anything cooled to the temperature of liquid nitrogen to come into contact with any part of the body since it can cause immediate and severe frostbite. Wear gloves and safety glasses when working with liquid nitrogen. Rare-earth magnets break easily, especially if they snap together. They can snap together with sufficient force to pinch your fingers or skin. When they break, they can shatter and cut you. Strong magnets can destroy computer disks and credit cards, harm computer monitors, and potentially affect heart pacemakers.

REFERENCES

1. P. J. Ouesph, *Phys. Teach.* **28**, 205 (1990).
2. M. K. Wu, L. R. Ashburn, C. J. Torng, P. H. Hor, R. L. Meng, L. Gao, Z. J. Huang, Y. Q. Wang, and C. W. Chu, *Phys. Rev. Lett.* **58**, 908 (1987).
3. C. W. Chu, L. Gao, F. Chen, Z. J. Huang, R. L. Huang, R. L. Meng, and Y. Y. Xue, *Nature* **365**, 323 (1993).
4. W. M. Saslow, *Am. J. Phys*. **59**, 1 (1991).
5. R. Brown, *Phys. Teach.* **38**, 168 (2000).
6. J. Bransky, *Phys. Teach*. **28**, 392 (1990).
7. A. M. Wolsky, R. F. Giese, and E. J. Daniels, *Scientific American* **260**, 60 (Feb 1989).
8. P. Weiss, *Science News* **158**, 330 (2000).

6
Light

As with electricity and magnetism, the study of light developed independently until 1865, when James Clerk Maxwell (1831 – 1879), by a purely theoretical argument based on a missing symmetry in the known laws of electricity and magnetism, added a term to the equations and predicted a form of electromagnetic wave that travels with a speed exactly equal to the then-known speed of light (3×10^8 meters/second).[1] In one swoop he consolidated light, electricity, and magnetism, as well as other forms of electromagnetic waves, many of which were yet unknown. Although light is a form of wave, in many instances, its behavior can only be understood by resorting to a particle description, not unlike the description that prevailed from the time of the ancient Greeks up through the time of Isaac Newton (1642 – 1727). The modern theory is that light has a dual nature that requires a quantum mechanical description. Light is a particularly suitable choice for demonstrating electromagnetic waves because complicated instrumentation is unnecessary since our bodies come equipped with a pair of remarkably sensitive and versatile optical detectors.

[1] The meter is now defined such that the speed of light is *exactly* 299,792,458 meters per second.

Safety Considerations with Lasers

Lasers are especially useful sources of light for demonstrations because they are monochromatic, coherent, collimated, and intense. However, the use of lasers in optics demonstrations poses special hazards particularly to an unwary audience [1–6]. Since most such demonstrations use relatively low-power lasers operating in the visible range, the discussion will be restricted to them. High-power lasers and lasers operating outside the visible range pose additional hazards and should only be used by a laser expert. The greatest hazard from visible lasers is damage to the retina of the eye. When light focuses on the eye, the intensity on the retina is a factor 10^4 to 10^6 greater than the intensity on the pupil of the eye. The dangers include the production of visible lesions, permanent partial "bleaching" of the pigment for one particular light color, and complete blinding if the light hits the optical nerve. You should use special glasses or goggles to protect your eyes, but they are usually impractical in a demonstration before a large audience.

To estimate the danger, you must know the power or energy output P of the laser in watts or joules, the beam diameter D at the aperture of the laser in centimeters, and the beam divergence φ in radians. The intensity of radiation at a distance d (in centimeters) from the laser is then calculated from

$$I = 4P / \pi(D+d\varphi)^2$$

in watts per square centimeter if P is the laser power, or joules per square centimeter if P is the laser energy. Compare this value with the damage threshold values in table 6.1. Note that even small demonstration lasers can cause eye damage if the beam strikes the eye directly. The laser light can usually be safely viewed after diffuse reflection (such as from a piece of paper), but you must avoid specular (mirror-like) reflections. Note also that an individual laser may exceed the nominal specifications given by the manufacturer.

To enhance safety, never allow the laser to operate unattended. Keep the illumination in the room as bright as possible to constrict the pupils of the observers' eyes. Avoid placing the laser so that the beam is at normal eye level. Use a diffuse, absorbing, fire-resistant target. Remove all watches and rings and other shiny objects near the laser. You can use lenses to defocus the beam, and you can place shields to prevent the beam from going beyond the area needed for the demonstration. In general, regard the laser as an ordnance piece with similar security and safety measures. With these precautions, the laser permits many impressive visual demonstrations.

REFERENCES

1. E. A. Lacy, *Handbook of Electronic Safety Procedures*, Prentice-Hall: Englewood Cliffs, NJ (1977).

2. D. Sliney and M. L. Wolbarsht, *Safety with Lasers and Other Optical Sources: A Comprehensive Handbook*, Plenum Press: New York (1980).

3. D. C. Winburn, *Practical Laser Safety*, Marcel Dekker: New York (1989).

4. L. L. Davey, Ed., *Code of Federal Regulations, Food and Drugs, Title 21*, Office of the Federal Register, National Archives and Records Administration, U. S. Government Printing Office: Washington (1985).

5. A. R. Henderson, *A Guide to Laser Safety*, Chapman & Hall: London (1997).

6. R. Henderson and K. Schulmeister, *Laser Safety*, Institute of Physics: London (2003).

Table 6.1

Approximate damage threshold values for retinal tissue for various continuous-wave (cw) and pulsed lasers in the visible (0.4–0.7 μm) range (minimum reported) [2]

Pulse length	Damage threshold intensity
cw[1]	10×10^{-3} W/cm^2
100 ns	$(100 \times 10^{-6}$ J/cm$^2)$[2]
30 ns	70×10^{-6} J/cm^2
20 ns	130×10^{-6} J/cm^2
1 ns	$(120 \times 10^{-6}$ J/cm$^2)$[3]
30 ps	18×10^{-6} J/cm^2

[1] Assumes a one-second maximum exposure.
[2] Estimate.
[3] Estimate.

6.1 Prism Rainbow

A rainbow produced by passing a collimated beam of white light through a glass prism illustrates that white light consists of many different colors.

MATERIALS

- slide projector or other collimated source of white light
- screen or light-colored wall
- glass prism[1] or bowl of water and mirror
- beaker of water and overhead projector (optional)
- slide with transparent slit (optional)
- laser (optional)
- spherical flask of water (optional)
- Variac (optional)
- large, clear light bulb (optional)
- second prism or magnifying glass (optional)

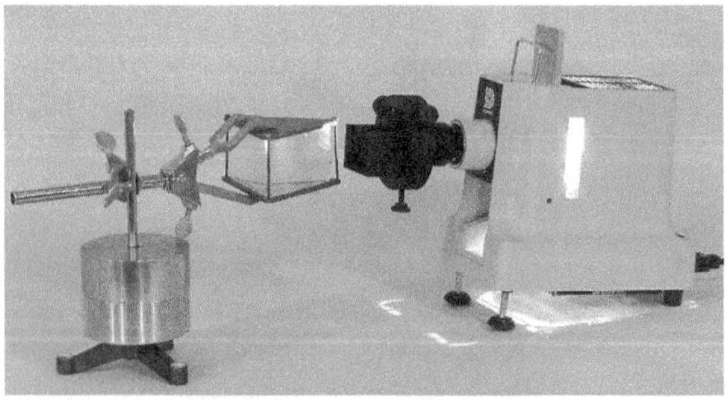

PROCEDURE

Pass white light from a slide projector or other bright source through a glass prism and project it on a screen or light-colored wall. You can use a bowl of water with a mirror immersed in the water or a beaker of water on an overhead projector instead if a prism is not available. The effect is best if the light is highly collimated and the room darkened. You can accomplish this by placing a slide with a transparent slit about a centimeter wide in the projector. Alternately, if the Sun is in the proper position, you can use light passing

---

[1] Available from American 3B Scientific, Carolina Biological Supply Company, Edmund Optics, Educational Innovations, Frey Scientific, PASCO Scientific, Sargent-Welch, Science Kits & Boreal Laboratories, and Scientifics.

through a window and a hole in a shade. You can use a laser to demonstrate that the effect occurs only for white light. To illustrate better the formation of rainbows by raindrops, use a spherical flask of water. You can use mirrors to rotate the spectrum into the usual position (arched in the center with red on the top). With a second prism or magnifying glass, you can show that the colors of the rainbow are not broken into further colors, but rather combine to form white. You can block out some of the colors in the space between the two prisms and observe the effect on the recombined light.

You can control the intensity of the lamp with a Variac to illustrate that at low temperature the light is mostly red but that the shorter wavelength colors successively appear as you increase the temperature of the lamp (Wein's law). Point out that there is radiation below the red (infrared) and above the violet (ultraviolet) but that our eyes are not sensitive to these wavelengths. With a detector sensitive to infrared,[2] you can demonstrate that energy is present beyond the visible region of the rainbow. You can also connect a large, clear light bulb to the Variac and show that it starts out glowing red at low current, and then becomes white as the other colors appear. Even before it is giving off significant light, it is still quite warm, indicating the emission of infrared radiation. Mention the analogy with ultrasound and infrasound, which are sounds too high and too low in frequency, respectively, for our ears to hear.

The band of wavelengths to which the eye is sensitive is the same as the wavelengths for which we receive the maximum radiation from the Sun. This apparent coincidence is probably a result of biological evolution. Approximately 70% of a person's sensory input is through vision [1]. Point out that red, green, and blue are the primary additive colors because they represent three roughly equidistant points on the spectrum from which the other colors can be approximately produced by mixing [2, 3]. Do not miss the chance to add appropriate music such as "Somewhere Over the Rainbow."

DISCUSSION

It was none other than Sir Isaac Newton (1642 – 1727) who first showed that white light contains all the colors of the rainbow, but he did not have a satisfactory explanation for the effect. The dispersion of white light into its constituent colors occurs because the index of refraction of water and most glasses varies with the wavelength of the light. Typically, the index of refraction is

$$n = A + B\lambda^2$$

called the "Cauchy formula," after Augustin Cauchy (1789 – 1857), where A and B are experimental constants that depend on the type of glass and λ is the wavelength in vacuum. Substances with a large value of B have the most dispersion. For most glasses, the index of refraction is about 1% greater at the short wavelength (violet) end than at the long wavelength (red) end of the visible spectrum. The result is to refract the violet light more at the interface between the air and the glass. The triangular shape of the prism causes the refraction at the entrance and exit interfaces to add rather than cancel as they would for a flat plate of glass.

---

[2] An ordinary silicon solar cell available from many sources is sensitive to infrared radiation.

Prism spectrometers are useful for determining the composition, temperature, and distance of stars. In the laboratory, spectrometers are used to identify the composition of materials, since each chemical element has its own characteristic spectrum of emission lines. Modern spectrometers are more likely to use diffraction gratings, rather than prisms, since they are more efficient, provide a linear dispersion, have less absorption of light, and have a wider useful range of wavelengths.

The refraction of white light reflected off the rear interior surface of raindrops produces rainbows in the sky [4–7]. About 4% of the light that attempts to exit a raindrop reflects internally, and red light travels about 1% faster in water than violet light does, accounting for the dispersion. A weak secondary rainbow occurs as a result of double reflection off the interior of the raindrops, and a tertiary rainbow occurs as a result of a triple reflection off the interior of the raindrops. The tertiary rainbow is on the opposite side of the main rainbow from the secondary rainbow and is hard to see because it is very dim and only 41° from the light source [4, 8–11]. It was René Descartes (1596 – 1650) who first correctly explained the mechanism that produces rainbows.

HAZARDS

Do not look directly into a bright source of light, but rather observe the spectrum projected on a screen or wall.

REFERENCES

1. P. Davidovits, *Physics in Biology and Medicine* (2nd ed.), Harcourt/Academic Press: San Diego (2001).
2. L. Parsons, *Phys. Teach.* **36**, 347 (1998).
3. L. D. Woolf, *Phys. Teach.* **37**, 204 (1999).
4. C. F. Boyer, *The Rainbow: From Myth to Mathematics*, Princeton University Press: Princeton, NJ (1987).
5. R. J. Whitaker, *Phys. Teach.* **12**, 283 (1974).
6. R. Greenler, *Rainbows, Halos, and Glories*, Cambridge University Press: Cambridge (1980), reissued by Peanut Butter Publishing: Milwaukee (1999).
7. H. M. Nussenzveig, *Scientific American* **236**, 116 (Apr 1977).
8. J. D. Walker, *Am. J. Phys.* **44**, 421 (1976).
9. J. D. Walker, *Scientific American* **237**, 138 (Jul 1977).
10. R. L. Lee and A. B. Fraser, *The Rainbow Bridge: Rainbows in Art, Myth, and Science*, Penn State University Press: University Park, PA (2001).
11. A. W. Hendry, *Phys. Teach.* **41**, 460 (2003).

6.2 Laser Beam

The beam from a low-power laser permits a number of simple demonstrations.

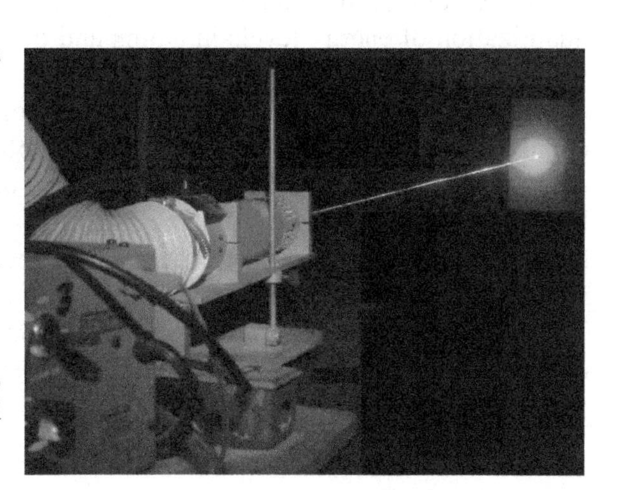

MATERIALS

- low-power, continuous, visible laser[1] or laser pointer
- chalk-board erasers, incense stick, ultrasonic humidifier, or liquid nitrogen
- smoke generator[2] (optional)
- glass prism (optional)
- diffraction gratings or fine-mesh screen (optional)
- Styrofoam® drinking cup and sponge (optional)

PROCEDURE

Use a low-power (few milliwatt), continuous, visible laser (helium-neon or equivalent) to project a beam across the front of the room. For a small audience, a laser pointer will suffice. Darken the room to render the beam visible. You can make the beam much brighter by clapping together two chalk-board erasers covered with chalk dust, or by using an incense stick [1], or a commercial smoke generator, or an ultrasonic humidifier of the type commonly used in homes [2], or a homemade generator produced with boiling liquid nitrogen [3].

There are many possible uses for such a laser beam [4–6]. You can illustrate the monochromatic nature of the laser light by passing it through a glass prism and comparing the result with the result of passing a collimated beam of white light through the same prism. You can scatter the beam off various objects. A fine-mesh screen or diffraction grating produces an interesting interference pattern when illuminated with a laser. You can pass the beam over a knife-edge, through various fluids and vapors, or through lenses of various types. Reflect the laser beam off a small mirror mounted on a loudspeaker connected to a source of sound. With two such loudspeakers, you can

---

[1] Available from American 3B Scientific, Carolina Biological Supply Company, Edmund Optics, Frey Scientific, Industrial Fiber Optics, PASCO Scientific, Sargent-Welch, and Ward's Science.

[2] Available from Scientifics.

construct a crude, low-frequency oscilloscope and display Lissajous patterns [7–13] by connecting the loudspeakers to sinusoidal sources with harmonically related frequencies.

With a diffraction grating, you can show that the laser light consists of a small number of distinct colors, as opposed to the light from an incandescent bulb that produces all the colors of the rainbow. Comment that this line spectrum is evidence of the quantization of energy levels in atoms and that it was a great puzzle to scientists when it was first discovered.

For a sophisticated audience, try the following puzzling demonstration. Pour an equal amount of water into each of two Styrofoam® drinking cups. Aim the laser at one of the cups for a few seconds. Turn the cups upside down. Water comes out of one cup, but little or none comes from the cup that you illuminated with the laser. A sponge surreptitiously wedged in the bottom of the cup before the demonstration soaked up the water. Ask who knew a laser would do that, and you will get many positive responses. This trick does not help one understand laser light, but it does give you a good opportunity to discuss the importance of careful observation and controlling all the variables in an experiment.

DISCUSSION

Laser light is intense, collimated, monochromatic, and coherent. A laser beam traveling through clear air is essentially invisible because there is little scattering of the light off air molecules. Chalk dust and smoke provide a multitude of tiny particles from which the laser beam can scatter. Since the laser beam is monochromatic, it deflects through a constant angle by a prism and is not dispersed. Coherence means that the photons of the light are all vibrating in synchronism with the same phase rather than with random phases as in ordinary light, and this coherence is a property of stimulated emission in which each photon produces a second identical one. The beam is collimated because the light reflects between two parallel mirrors at the end of a long tube inside the laser, with one mirror only partially reflecting to allow some of the light to escape. By intentionally misaligning the mirrors, you can often produce light whose intensity fluctuates chaotically.

The monochromatic and coherent nature of the beam makes it ideal for interference and diffraction experiments. Lasers are used in CD and DVD players, in barcode scanners, in laser printers, and in light pointers for lectures. The police use laser guns to check traffic speeds. Laser light is also used to make holograms. The word "laser" is an acronym for "light amplification by stimulated emission of radiation." Albert Einstein (1870 – 1955) postulated the existence of stimulated emission.

HAZARDS

The only significant danger from a laser of this class is eye damage from looking directly into the beam. Take care in setting up the demonstration to ensure that the laser beam cannot be directed or reflected into someone's eye.

REFERENCES

1. R. Ebert, *Phys. Teach.* **30**, 185 (1992).
2. D. Heiden, *Phys. Teach.* **35**, 94 (1997).
3. M. E. Knotts, *Phys. Teach.* **31**, 402 (1993).
4. H. H. Gottlieb, Ed., *101 Ways to Use a Laser*, Metrologic Instruments Inc.: Bellmawr, NJ (1984).
5. E. Schmidt, *Phys. Teach.* **27**, 30 (1989).
6. J. O'Connell, *Phys. Teach.* **37**, 445 (1999).
7. T. Campbell, *Phys. Teach.* **10**, 283 (1972).
8. K. D. Pinkerton, *Phys. Teach.* **29**, 168 (1991).
9. T. B. Greenslade, *Phys. Teach.* **31**, 364 (1993).
10. G. Schuttinger, *Phys. Teach.* **31**, 375 (1993).
11. E. Y. C. Tong, *Phys. Teach.* **35**, 491 (1997).
12. T. B. Greenslade, *Phys. Teach.* **41**, 351 (2003).
13. C. Criado and N. Alamo, *Phys. Teach.* **42**, 248 (2004).

6.3
Laser Gun

A sufficiently powerful laser bursts balloons from across the room.

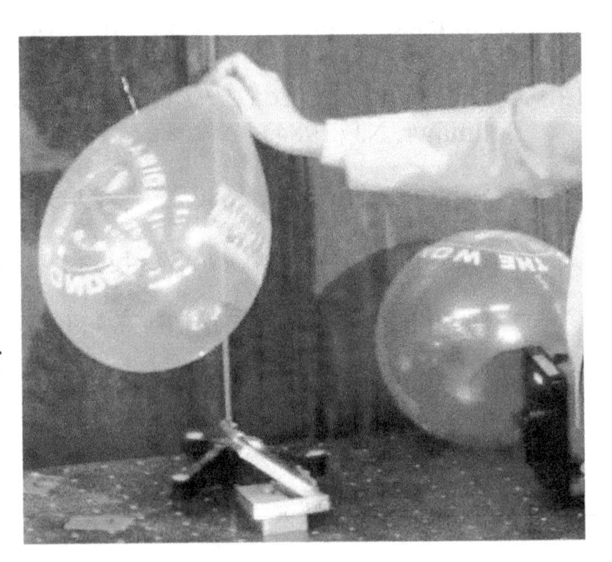

MATERIALS

- high-power, continuous laser (at least one watt average power)
- assorted colored balloons and pieces of colored paper
- beam dump
- optical lenses (optional)

PROCEDURE

If a sufficiently powerful laser is available, you can perform spectacular demonstrations if you are careful to avoid the hazards inherent with intense light sources (see the section on Laser Safety Considerations). A laser of at least a watt average power is desirable. Contrast the power used here with that used in more traditional lasers (a few milliwatts) as well as ordinary incandescent light bulbs (100 watts or so). Pulsed lasers with high peak power and lasers outside the visible range also work for this purpose but are more dangerous and are not recommended. You should position the beam line to avoid shining the laser into someone's eye, and follow the usual precautions of avoiding specular reflections. Terminate the laser beam on a beam dump capable of dissipating all the power of the laser without significant reflection or overheating. A long, corrugated metal pipe, blackened on the inside and gently bent into a circular arc, makes a suitable beam dump. Reduce the background room light to make the laser beam more visible, but beware that the eye is more sensitive and susceptible to damage in low illumination.

If the laser is of marginal power, try focusing the laser beam with a pair of lenses. Place a diverging lens with a long focal length in the beam. Then bring the diverging beam to a focus using a converging lens with a short focal length. Not only does this make an intense spot, but it defocuses the beam and makes it safer for the audience.

With a laser of this class, it is possible to burn holes and ignite pieces of paper and to burst balloons. A light-colored paper or balloon reflects most of the light and is more difficult to burn or burst than one with a dark color, which efficiently absorbs the laser light. A particularly nice demonstration is to enclose a dark-colored balloon inside a clear one. You can make the laser beam pass though the clear balloon and burst the inner balloon. Balloons filled with flammable gas (methane, hydrogen, etc.) cannot be ignited directly because such gases are transparent to the laser beam, but with an appropriate piece of paper taped to the balloon, you can set the paper on fire, and ignite the gas indirectly. When you have the clear balloon, you could offer to pop it with a needle, which you then pass though the balloon at places that have small patches of transparent tape that prevent the balloon from popping. You can also pass the needle though the thick layer of rubber adjacent to the neck and directly opposite the neck without popping it [1, 2]. Then remove the needle and pop the balloon in the regular way. Explain how you are able to do this trick. You can also pass a pencil through a Ziploc® bag filled with water in the same way [1].

On a sunny day you can similarly use a magnifying glass outdoors to concentrate light to the point where it ignites a target. With a sufficiently bright incandescent or arc lamp and an appropriate lens, you can do the same in the lecture room to illustrate that there is nothing special about using a laser other than its natural collimation.

DISCUSSION

Although a watt is a relatively small amount of power, the laser has the ability to concentrate this power in a spot of only a few millimeters in diameter. If the surface on which the laser beam is incident absorbs the light efficiently enough, the heat can puncture or ignite the material. The color and to a lesser extent the texture of the target determines how much of the light is absorbed. Light not absorbed is either reflected or transmitted through the target. When performing this demonstration, it is difficult to avoid implications about the prospect of directed energy weapons [3, 4]. Although it appears to illustrate the feasibility of such applications, you can equally well use it to illustrate the ease with which one can make a target that absorbs very little of the laser light or even reflect it back to the source.

HAZARDS

This demonstration is potentially very dangerous. If the light from a laser of this class impinges on one's eye, it can cause almost instant blindness. Only someone experienced with the use of high-power lasers should do this demonstration and only after taking the necessary precautions to ensure that the laser beam cannot reflect into someone's eye. The laser is also capable of burning one's skin and setting fire to flammable materials placed in its path.

REFERENCES

1. R. D. Edge, *Phys. Teach.* **30**, 379 (1992).
2. M. Gardner, *Phys. Teach.* **34**, 233 (1996).
3. K. B. Payne, *Laser Weapons in Space: Policy and Doctrine*, Westview Press: Boulder, CO (1983).
4. N. Bloembergen and C. K. Patel, *Rev. Mod. Phys*. **59**, 3, part II (1987).

6.4
Twinkling Stars

A laser beam passed over the top of a Bunsen burner produces a spot on the wall that twinkles like a star.

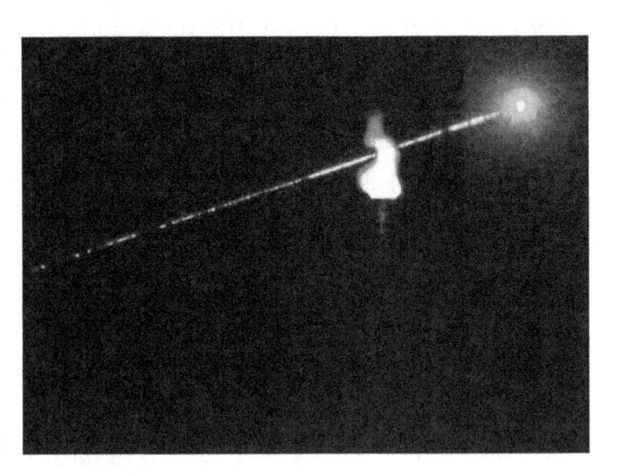

MATERIALS

- low-power, continuous, visible laser
- Bunsen burner
- slide projector and photographic slide of stars (optional)

PROCEDURE

Position a laser so that the beam passes over the top of a Bunsen burner and makes a spot on the wall some distance away. After pointing out the spot on the wall to the audience, light the Bunsen burner. The spot begins to dance around like a twinkling star. Then extinguish the Bunsen burner to make the spot stop moving.

Alternately, in place of the laser, use a slide projector and a photographic slide of stars. Lacking a suitable slide, you can make one with aluminum foil with a number of pinholes perhaps arranged in the form of the Big Dipper or some other easily recognized constellation.

Do not miss the chance to play that famous composition by Mozart that everyone knows as "Twinkle, Twinkle, Little Star."[1]

DISCUSSION

Light travels in a straight line only when traveling in a medium with a constant index of refraction. The index of refraction of air depends slightly on its density. At a given pressure, the density of a gas is inversely proportional to its absolute temperature. Thus the hot air above the flame of the Bunsen burner deflects the light slightly, and in a time-varying way, due to the turbulence of the air above the flame. The same phenomenon occurs when you view the stars through the turbulent atmosphere of the Earth. In that case the temperature gradients are smaller, but the path length is much longer.

The myth abounds that stars twinkle, but not the planets. If this is so, it is only because on average the planets appear larger (although dimmer because their light is

---

[1] Actually, Mozart did not write it. The tune in his day was known as "Ah, vous dirais-je, maman," and he wrote a set of twelve variations on it (K. 265).

reflected) than stars, which are essentially point sources, and thus the effect is less noticeable [1]. The apparent position of the Moon presumably dances around in a similar way, but it occupies such a large segment of the sky that we never notice it.

This phenomenon is what makes it ineffective simply to build ever-larger telescopes to improve the resolution of astronomical objects. Satellite-based telescopes or telescopes positioned on the Moon do benefit from a larger size, however.

HAZARDS

You should observe the usual precautions concerning the use of lasers. Set up the demonstration so that there is no chance of the laser beam directly entering anyone's eye. The Bunsen burner is an obvious source of fire and burns, especially if you do the demonstration in subdued illumination.

REFERENCE

1. J. S. Huebner and T. L. Smith, *Phys. Teach.* **32**, 102 (1994).

6.5
Spiral Light Guide

A long, solid, plastic, spiral rod illuminated with a low-power laser illustrates the property of total internal reflection in a light guide.

MATERIALS

- 1/2-inch-diameter solid rod of Plexiglas® or other plastic[1]
- low-power, continuous, visible laser
- LED, fiber optics, and solar cell (optional)
- waterproof flashlight or laser pointer and aquarium filled with water (optional)

PROCEDURE

Heat the rod and bend it into a spiral or other shape, and polish the ends. Illuminate one end of the rod with the laser. The entire rod lights up along its length due to scattering off imperfections in the rod. A bright spot appears at the opposite end of the spiral. The effect is best when viewed in subdued background light. Such a configuration illustrates the principle of the light guide (also called a "light pipe"). Emphasize that the rod is solid, not a hollow tube. With a smoked Lucite® rod,[2] you can more easily see the path of the light inside the rod.

You can show other examples of light guides, especially the flexible kind made of many fine strands (typically 10 μm in diameter) of silica glass (fiber optics). With modern, low-loss fibers, light will travel for many kilometers with little attenuation [1–5]. Attenuation in modern fibers is less than 1 dB/km. Such technology is revolutionizing communications because the high-frequency light provides a transmission

[1] Available from Frey Scientific, Sargent-Welch, and Scientifics.
[2] Available from Arbor Scientific.

medium with extraordinarily large bandwidth. Data rates as high as 27 gigabits per second are possible. This rate is equivalent to four hundred thousand telephone conversations. You can use a waterproof flashlight or laser pointer immersed in a water-filled aquarium to illustrate total internal reflection more directly.

With an LED or laser diode or even an incandescent bulb in place of the loudspeaker of a small transistor radio, a length of fiber optic light guide, and a solar cell or photo resistor connected to an amplifier and loudspeaker, you can demonstrate the transmission of audio using fiber optics. In a more elaborate setup, you can even multiplex two audio signals [6].

By maintaining the relative position of the many strands within the bundle (a coherent fiber bundle), it is possible to pipe an image from place to place. Such devices are called fiberscopes and are used to see places that are inaccessible to the eye, such as inside someone's stomach, heart, or bowels, in which case they are called endoscopes. Most fiberscopes consist of two fiber bundles, one to provide illumination from a high-intensity source such as a xenon arc lamp, and the other to view the image. In the usual arrangement, the illumination bundle surrounds the light-collecting bundle.

DISCUSSION

The light guide illustrates the principle of total internal reflection. According to Snell's law, light traveling in a medium with index of refraction n_1 and incident on a material with index of refraction n_2 at an angle θ_1 with respect to the normal at the interface refracts at an angle θ_2 with respect to the normal such that $n_1 \sin\theta_1 = n_2 \sin\theta_2$. If n_1 is greater than n_2, then there is a critical angle $\theta_1 = \theta_c$ such that $\theta_2 = 90°$. Light incident on an interface with $\theta_1 > \theta_c$ will experience total internal reflection. With $n_1 = 1.51$ for Plexiglas® and $n_2 = 1.0003$ for air, the critical angle is $\theta_c = \sin^{-1}(n_2/n_1) = 41.6°$. The condition for total internal reflection is easily met provided the light guide is long and skinny and not bent at too sharp of an angle. The critical angle for water ($n_1 = 1.33$) and air is about 49°.

The experimental discovery of Snell's law is usually credited to Willebrord van Roijen Snell (1580 – 1626), but it was also deduced from the particle theory of light by René Descartes (1596 – 1650) and is known as "Descartes' law" or the "law of sines" in France. Although Snell discovered the law in 1621, he did not publish it until after Descartes had done so in 1637, leading Christiaan Huygens (1629 – 1695) to accuse him of plagiarism. Some sources, such as Schribner's Dictionary of Scientific Biography, spell Snell's name with a single "l," but Snell himself spelled it "Snellius," writing in Latin. Even earlier, the English scientist Thomas Harriot (1560 – 1621) described the law of sines [7].

HAZARDS

This demonstration is safe provided you exercise normal caution to prevent the laser and its specular reflections from shining into anyone's eye.

REFERENCES

1. W. S. Boyle, *Scientific American* **237**, 40 (Aug 1977).
2. S. J. Buchsbaum, *Physics Today* **29**, 23 (May 1976).
3. E. A. Lacy, *Fiber Optics*, Prentice-Hall: Englewood Cliffs, NJ (1982).
4. E. E. Basch, Ed., *Optical-Fiber Transmission*, Howard W. Sams & Company: Indianapolis (1987).
5. C. D. Chaffee, *The Rewiring of America: The Fiber Optics Revolution*, Academic Press: San Diego (1987).
6. A. Niculescu, *Phys. Teach.* **40**, 347 (2002).
7. J. M. Dudley and A. M. Kwan, *Phys. Teach.* **35**, 158 (1997).

6.6 Water Light Guide

A stream of water illuminated with a laser or high-intensity white light acts as a light guide.

MATERIALS

- tank of water with a hole in the side[1]
- continuous, visible laser or bright light bulb
- trough or bucket for catching the water
- Fluorescein® or other fluorescent substance (optional)
- household ammonia and phenolphthalein indicator (optional)

PROCEDURE

Fill the tank with water that streams out a hole in the side [1, 2]. A carefully aimed laser that shines through a window opposite the hole through which the water streams provides the light. The window can consist of a piece of glass bonded with putty or bathtub cement over a hole in the tank. Alternately, you can lower a light bulb into the water. In the latter case, the bulb should be as bright as possible (a few hundred watts), and the inside of the tank should be white or shiny so as to reflect as much of the light as possible. Catch the water in a trough or bucket on the floor. Seal the tank tightly to prevent escape of the light, but there should be an opening near the top to allow air to enter as the water leaves. You can best illustrate the effect in a darkened room. You can add Fluorescein® or other fluorescent substance to the water for increased visibility. The tank should contain enough water to provide a nearly constant flow for about a minute (a few gallons), and the height of the water column should be a good fraction of a meter to provide sufficient pressure at the aperture as required for a steady stream of water. You can hold your finger or hand in the water stream to illuminate it and to show that the stream consists of water since it otherwise looks quite rigid. As the level of water in the tank decreases, the stream bends at a sharper angle until suddenly the condition for total internal reflection is violated, and a bright spot appears on the opposite wall.

[1] Available from Sargent-Welch.

A more modest demonstration can be done using a 2-liter clear plastic soda bottle with a small (1/8-inch-diameter) hole about 5 cm from the bottom. An easy way to drill a clean hole in a plastic bottle is to first fill it with water and then freeze the water.

A clever variant of the demonstration involves adding a bit of household ammonia to the water and some phenolphthalein (the second "ph" is silent) indicator in the catch trough with a red laser [3]. After doing the demonstration in the dark, ask the audience if they have heard of the "conservation of color." Then turn on the lights, and show them that the water in the basin is red. After letting them puzzle over that for a moment, explain that the catch trough contained a chemical indicator that turns red in the presence of the hydroxide (OH^-) ions present in the ammonia water. If you do not have access to phenolphthalein, you can make some by crushing an Ex-Lax® tablet in 30 milliliters of rubbing alcohol [4]. As an alternative, put a small red light in the bucket and claim that the water trapped the red light.

DISCUSSION

The index of refraction of water is 1.33, and thus the critical angle for a water/air interface is $\sin^{-1}(1/1.33) = 49°$. As long as the water stream does not bend at too sharp an angle, light traveling along the length of the stream strikes the water/air interface at an angle greater than 49° with respect to the normal to the interface and is thus totally reflected. Many lighted fountains exhibit this phenomenon.

Unlike most of the demonstrations in this book, the inventor and the time and place of its first public display are known for this demonstration, which is sometimes called "the liquid vein." It was invented by John Tyndall (1820 – 1893), professor of natural philosophy at the Royal Institution in London, and presented in his Friday Discourse "On some Phenomena connected with the Motion of Liquids" at the Royal Institution on May 19, 1854 [5]. It was also used by John Henry Pepper (1821 – 1900), a chemist at the London Polytechnic Institution, in the following years [6].

HAZARDS

If a laser is used, take care to ensure that it does not shine into someone's eye. If you immerse a light bulb in the water, it is best to use a low-voltage bulb, such as an old 12-volt automobile headlamp, to minimize the danger of electrocution. Immerse the lamp in the water before you turn it on to prevent the thermal shock from breaking it. Aim the water stream so that it will not damage other equipment, cause an electrical hazard, or make the floor slippery. Promptly clean up any water that spills.

REFERENCES

1. H. A. Robinson, Ed., *Lecture Demonstrations in Physics*, American Institute of Physics: New York (1963).
2. A. Kshatriya, *Am. J. Phys.* **44**, 604 (1976).

3. G. R. Gore, *Phys. Teach.* **30**, 352 (1992).

4. A. S. W. Sae, *Chemical Magic from the Grocery Store*, Kendall/Hunt: Dubuque, IA (1996).

5. T. B. Greenslade, *Phys. Teach.* **35**, 207 (1997).

6. J. H. Pepper, *The Boy's Playbook of Science* (2nd ed.), Routledge, Warne, and Routledge: London (1860).

6.7 Rayleigh Scattering

A white light passing through a liquid scatters primarily the blue light causing the transmitted light to appear red, simulating the blue sky and setting Sun.

MATERIALS

- overhead projector and large glass beaker with a piece of cardboard having a circular hole, or
- slide projector (or bright flashlight) and aquarium
- small amount of skim milk or coffee-whitener or a solution of thiosulfate and hydrochloric acid
- stirring stick
- precious opal crystal (optional)
- Polaroid® sheet[1] (optional)
- laser (optional)

PROCEDURE

There are two versions of this demonstration (also called "Tyndall's experiment"), one using an overhead projector and a large glass beaker and the other using a slide projector and an aquarium [1–5]. In the first version, place on the overhead projector a piece of cardboard with a circular hole about half the diameter of the beaker (so as to allow multiple scattering of the light by the water), and set the beaker filled with water over the hole. Form a circular image from the projector on the screen to simulate the Sun.

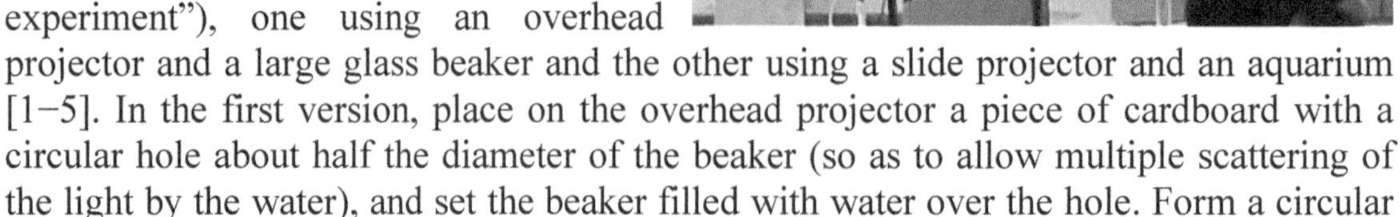

[1] Available from Edmund Optics, Educational Innovations, PASCO Scientific, and Scientifics.

In the other version of the demonstration, adjust the slide projector to produce a beam of light so that it makes an image resembling the Sun on a screen. You may need to insert a slide with a circular aperture to make a good image. You can use a bright flashlight in place of the slide projector, although the image will then be more diffuse. Shine the light through the aquarium filled with water.

Then add a small amount of skim milk or coffee-whitener to the water while stirring with a stick. If you have easy access to chemicals, use a mixture of 100 ml of 5% sodium thiosulfate solution and 100 ml of 0.05 M hydrochloric acid in place of the milk. The reaction slowly produces a colloidal solution of sulfur, causing the "Sun" to set slowly over the course of several minutes. Alternately, use a mixture of cedarwood oil and alcohol. You can replace the liquid with a crystal of precious opal, which you can purchase from many sources for just a few dollars [6].

The scattered light will look decidedly blue, and the transmitted light, which is deficient in blue light, will look somewhat red. Explain that the scattering of light off particles in the air is what makes the sky blue and the sunset or sunrise red. Most of the blue light scatters out before the light from the setting Sun reaches your eye since it must travel though a considerable distance in the dense atmosphere. Sunsets are especially colorful when enhanced by air pollution, forest fires, or volcanic eruptions. Use a Polaroid® sheet to show that the scattered light is polarized.

The same kind of scattering off tobacco smoke gives it a blue color. In either case, it is important to add the milk slowly since too much milk will scatter all the light, making it look white, and there will be no transmitted light, or the image will be too dim. The demonstration is most effective in subdued room light. You can demonstrate that the phenomenon is absent when you use a monochromatic beam of laser light.

DISCUSSION

When light scatters off particles whose size is small compared to the wavelength of the light, the short wavelengths are scattered more than the long wavelengths. This phenomenon is called Rayleigh scattering after Lord Rayleigh, also known as John William Strutt (1842 – 1919), who showed that the scattering cross section in such a case is proportional to $1/\lambda^4$, where λ is the wavelength of the light. The formula for the intensity I of Rayleigh scattered light is

$$I/I_0 = 8\pi^4 N\alpha^2(1 + \cos^2\theta) / \lambda^4 R^2$$

where I_0 is the intensity of incident light, N is the number of scatterers, θ is the scattering angle, and R is the distance from the scatterers. Violet light is scattered even more than blue light, but our eyes are less sensitive to violet than to blue.

When the particles are larger than the wavelength, all wavelengths are scattered similarly, and the phenomenon is called Mie scattering after Gustav Mie (1868 – 1957). Clouds and fog are white because they contain small droplets of liquid water, typically a thousandth of an inch in diameter, which is much larger than the wavelength of light (several hundred nanometers). When the scattered light shifts frequency, the phenomenon is called Raman scattering after Sir Chandrasekhara Venkata Raman (1888 – 1970).

The night sky is also blue, as you can demonstrate by taking a time-exposed photograph [7]. The ocean appears blue in part because blue light from the sky illuminates it. Mars appears red because of dust in its atmosphere and because of the color of its rocky surface.

HAZARDS

The hazards of this demonstration are minimal. Do not shine a bright light into anyone's eye, and be careful not to break the beaker or aquarium or to spill water on the floor, which might cause someone to slip. If you use hydrochloric acid in place of milk, keep it away from contact with your skin.

REFERENCES

1. R. M. Sutton, *Demonstration Experiments in Physics*, McGraw-Hill: New York (1938).
2. H. Kruglak, *Phys. Teach.* **11**, 559 (1973).
3. M. H. Moore, *Phys. Teach.* **12**, 436 (1974).
4. C. F. Bohren and A. B. Fraser, *Phys. Teach.* **23**, 257 (1985).
5. J. S. Huebner, *Phys. Teach.* **32**, 147 (1994).
6. E. Zhu and S. Mak, *Phys. Teach.* **32**, 420 (1994).
7. R. Greenler, *Rainbows, Halos, and Glories*, Cambridge University Press: Cambridge (1980), reissued by Peanut Butter Publishing: Milwaukee (1999).

6.8 Fluorescence

Materials illuminated with ultraviolet light emit visible light.

MATERIALS

- ultraviolet lamp
- fluorescent materials
- Fluorescein® disodium salt[1] (optional)
- 0.8% aqueous solution of polyethylene oxide[2] (optional)
- phosphorescent sheet[3] (optional)

PROCEDURE

Motivate this demonstration by showing a rainbow spectrum of white light, and then introduce the idea of invisible light such as infrared and ultraviolet. Darken the room lights, and turn on an ultraviolet lamp. You can purchase ultraviolet lamps from most vendors of scientific supplies or from florists or hardware stores. Place various minerals and other fluorescent materials under the lamp. Some articles of clothing fluoresce nicely. Common household products that fluoresce are Murine® eye drops and Pearl Drops® toothpaste. Try brushing your teeth with such toothpaste, and continue lecturing with glowing bright yellow teeth. The coating on the inside of a fluorescent lamp works well except that the ultraviolet light does not easily penetrate the glass. Breaking such tubes is dangerous since the fluorescent material and the mercury in such tubes are highly toxic. Fluorescent lamps[4] are themselves familiar examples of fluorescence and can be adapted to a variety of uses including the demonstration of plasma effects [1] (see section 4.8).

A nice demonstration is to add about 10 mg of Fluorescein® ($Na_2C_{20}H_{10}O_5$) to 500 ml of a solution of 0.8% polyethylene oxide (Polyox WSR-301). Polyethylene oxide (long chains of $-CH_2-CH_2-O-$) is a viscoelastic liquid with many strange and delightful properties such as the ability to self-siphon [2]. With a pair of 1-liter beakers, pour the liquid from one beaker to the other. When you stop pouring it, the liquid continues to run

[1] Available from Aldrich Chemical Company, 1001 W. Saint Paul Avenue, Milwaukee, WI 53233, 414-273-3850, http://www.sigmaaldrich.com/.
[2] Available from Educational Innovations.
[3] Available from Educational Innovations.
[4] Fluorescent lamps coated on only one end are available from Science Kits & Boreal Laboratories.

up and over the rim of the upper beaker in seeming defiance of gravity. Cut the stream of liquid with a pair of scissors, and the upper part will jump back into the upper beaker. The Fluorescein makes it visible under ultraviolet light and offers a striking visual effect.

DISCUSSION

Fluorescence is the process whereby a substance absorbs light at a short wavelength and then re-emits it at a longer wavelength. All that is required is appropriately spaced energy levels in the atoms that constitute the material. The fluorescent lamp is the most common example. In such a lamp, electrons produced by a filament at one end of the tube travel down the tube and excite atoms of mercury vapor in the tube. The mercury emits ultraviolet light, which strikes the fluorescent coating on the inside of the glass.

The atoms of a fluorescent material typically re-emit their radiation after about 10^{-8} seconds. By contrast, phosphorescent materials continue to emit after you remove the incident radiation, sometimes for hours. Such materials are used in paints and on the hands of watches and clocks. Other light-emitting processes are chemiluminescence, in which a chemical reaction produces the light, and triboluminescence, in which mechanical stress in a crystal produces the light.

Viscoelastic liquids contain molecules with long chains of atoms, numbering into the millions. If you could view them under enormous magnification, they would look like a plate of spaghetti. With the molecules sufficiently entangled, they exhibit both high viscosity and elasticity. Exercise care in handling these materials since heat or vigorous stirring can break the molecules and degrade the effect. They should be prepared within a day of use and stored in a refrigerator. Viscoelastic materials (also called Memory Foam®) are used in mattresses and shoes to conform to the shape of one's body. Like plastic, these materials have properties of both liquids and solids, in that they flow like a liquid (with viscosity) but tend to return to their original shape (elasticity).

HAZARDS

Intense ultraviolet light can damage the eyes. Avoid prolonged staring at such lamps. In extreme cases, other parts of the body can be sunburned.

REFERENCES

1. N. R. Guilbert, *Phys. Teach.* **34**, 20 (1996).

2. A. A. Collyer, *Physics Education* **8**, 111 (1973).

6.9 Talking Head

Reflections from a mirror mounted beneath a table give the illusion that a disembodied head is sitting on top of the table.

MATERIALS

- table, about 1-meter-square with a circular hole near the back
- mirror to fit diagonally between opposite legs of the table
- tablecloth reaching to the floor on all sides
- shag carpet or straw (optional)

PROCEDURE

Mount the mirror underneath the table so that it covers the whole space from the underside of the tabletop to the floor [1]. A shag carpet or straw on the floor helps conceal the bottom edge of the mirror. Near the back of the table in the center and behind the mirror is a circular hole through the tabletop of sufficient diameter to pass the head of the volunteer. Cover the table initially with a cloth that reaches to the floor on all sides.

Take a volunteer from the audience around to the back of the table perhaps using an assistant who helps the person kneel or sit with the head protruding through the hole in the table while you, acting as a magician, hold the tablecloth in such a way as to conceal from the audience what is happening to the volunteer. A little music and patter about the similarities and differences between science and magic fits in well here. A few grunts and groans from behind the cloth add to the drama. Finally, when the volunteer is in place, remove the cloth and present the audience with the illusion of a disembodied head resting on the table. Carry on a conversation with the head, culminating perhaps in a pun about restoring the body while the volunteer is "ahead." In a magic show, you would remove the volunteer from the table out of view, leaving the audience to wonder how you did the illusion. However, in a science demonstration, you should remove the volunteer in view of the audience while explaining the trick. Point out that some knowledge of physics is essential to a good magician. A good joke is to say that magicians never explain their tricks, but physicists will just not shut up about them.

Whenever you use a volunteer from the audience, take a few moments to become acquainted with the person. Ask the person's name, whether they like science, what they study, what they want to be when they get older, and something relevant to the demonstration in which they are about to participate. In a case like this, ask if they enjoy magic and if they have ever been to a magic show. Ask if they have ever seen a magician remove someone's head (Woody Allen's second favorite organ).

The illusion is most effective if the mirror is very clean and has concealed edges. You should place the table well away from other obstructions that the mirror would obscure. If a carpet is used, align it carefully so that there is no discontinuity at its edge where it goes behind the mirror. Be careful not to stand directly in front of the table lest the reflection of your legs appear in the mirror. Do not place the table so close to the audience that they can see their reflection in the mirror. The illusion works best if the audience is seated slightly above the level of the top of the table. The proximity of the audience to the head would seem to favor discovery of the trick, but, on the contrary, it is indispensable to its success.

DISCUSSION

Although this demonstration is more amusing than educational, it serves to introduce, motivate, and illustrate the idea that the angle of incidence is equal to the angle of reflection in geometric optics, and it illustrates a virtual image. It also illustrates the danger of being deceived while making observations of nature and of the importance of considering all possible explanations of a phenomenon before reaching a conclusion.

HAZARDS

There are no significant hazards with this demonstration. Take care when moving the mirror to avoid breaking it, and caution the volunteer not to lean or push against the mirror.

REFERENCE

1. D. H. Charney, *Magic*, Strawberry Hill Publishing Company: New York (1975).

6.10
Pepper's Ghost

An illusion in which a person or other object disappears and reappears illustrates the phenomenon of the partial reflection of light at the interface between two media.

MATERIALS

- large plate of glass
- two spotlights on dimmer controls
- candle and beaker of water (optional)
- skeleton[1] or ghost (optional)

PROCEDURE

The simplest demonstration consists of a candle and a beaker of water [1]. Place the beaker of water behind the glass and the candle an equal distance in front of the glass so that the audience sees the beaker through the glass and sees the reflection of the candle in the glass. It looks as though the candle is burning under the water. A black shield between the candle and the audience prevents the audience from seeing the candle directly.

A more elaborate demonstration [2, 3] requires a much larger piece of glass (at least five feet square). Place the glass at a 45° angle to the audience so that the audience sees a combination of light passing through from behind the glass and light reflecting off the glass at a 90° angle from the line of sight. Spotlights on dimmer controls can alternately illuminate the area behind the glass or the area off to the side. Turn one light up while you turn the other down in such a way that the total light intensity is nearly constant. You can do this automatically with a single control or with practice by manipulating a pair of controls, one in each hand. You can also

[1] Available from Carolina Biological Supply Company and Scientifics.

just turn up the light that was originally off before turning down the one that was originally on. In this way, you can make people and other objects appear and disappear at will. It helps if the background behind both objects is dark and if you provide a partition to prevent the audience from directly seeing the object off to the side. Lower the room lights to prevent audience members from seeing their own reflection.

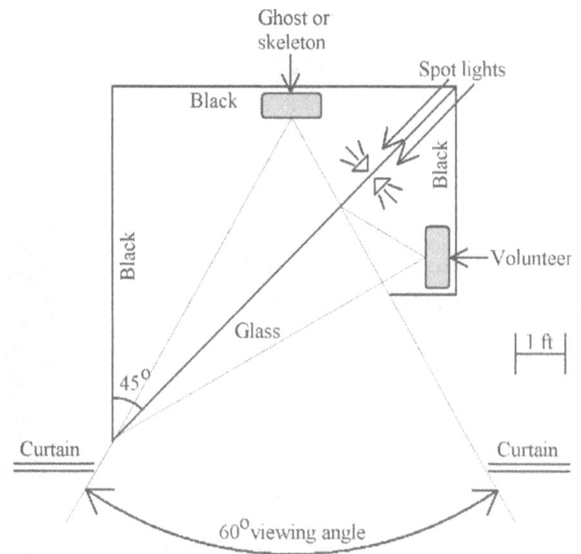

By proper manipulation of the lights, you can turn a volunteer into a ghost or skeleton and back again. It is important to place the person and ghost or skeleton in exactly the right position and to choose a volunteer who is nearly the same height as the ghost or skeleton (see figure). With proper placement, the figure will change continuously from one to the other independent of the viewer's position in the audience. In practice, you can accommodate a viewing angle of about 60°, depending on the width of the glass. This arrangement also makes an effective and dramatic entrance and exit for you. Glitter sprinkled from above in place of the skeleton aids in the appearance and disappearance and gives the appearance of a transporter from the television series *Star Trek.* Discuss the current impossibility of disassembling an object, transporting it electronically, and then reassembling it at a distant point. Point out that magicians often use illusions such as this and that some knowledge of physics is essential to a good magician. The illusion (also called "the blue room illusion") was originally devised and performed on stage in 1862 by John Henry Pepper (1821 – 1900), a professor of chemistry at London Polytechnic Institute.

DISCUSSION

A light beam incident on the interface between two transparent media of different indices of refraction (such as air and glass) partially reflects and partially transmits through the interface. Thus the light emitted from such an interface consists of the superposition of light traveling through the glass from behind and light reflected off the glass from a direction such that the angle of incidence is equal to the angle of reflection. By controlling the relative intensities of the two light sources, you can make one scene transform into another.

HAZARDS

There are no hazards with this demonstration other than the danger of breaking the glass or being cut by its sharp edges, which should be protected.

REFERENCES

1. D. H. Charney, *Magic*, Strawberry Hill Publishing Company: New York (1975).

2. J. P. VanCleave, *Teaching the Fun of Physics*, Prentice-Hall: New York (1985).

3. M. Pendergrast, *Mirror Mirror: A History of the Human Love Affair with Reflection*, Basic Books: New York (2003).

6.11
Tubeless Television

A visual image appears in midair when waving a light-colored stick near the focal plane of a slide projector containing a slide of some appropriate subject, illustrating the persistence of vision and the scanning process in television.

MATERIALS

- slide projector
- photographic slide of some subject
- light-colored stick

PROCEDURE

Aim the slide projector at a dark-colored, irregularly shaped background some distance behind you so that the image of the slide on the back wall is dim and out of focus. Alternately, project the image through an open door. Wave the stick so rapidly that one's persistence of vision produces a continuous image. The effect is best when viewed in subdued light. It is necessary to focus the projector at a distance such that the audience can see the image clearly from the back of the room and to wave the stick at the right place to obtain a well-focused image. You will need to practice getting the optimal distance and light level. Alternately, place the slide projector on the floor behind the lecture bench and aim it upward. A mirror at the back edge of the lecture bench then deflects the light horizontally so that the projector is not visible to the audience. You can use a vibrating string or rope in place of the stick. A nice touch is to use a picture of some well-known figure such as Albert Einstein.

You can use this demonstration as an introduction to the discussion of holography [1–13]. Holograms were invented by Dennis Gabor (1970 – 1979) at the Imperial College of London. In 1971, Gabor received the Nobel Prize in physics for holography. Holograms appear on credit cards, postage stamps, and driver's licenses, and other places. Someday, three-dimension holographic televisions will probably be available.

DISCUSSION

The physics principles involved are rather rudimentary and involve geometric optics and the physiological process of persistence of vision. Wave the stick at various rates to illustrate the extent of the visual persistence. The effect is analogous to the role of the persistence of vision in television and motion pictures and to the scanning process in the transmission of a television picture. The scanning of a television image can often be made apparent by waving your fingers rapidly up and down between your eyes and the television monitor. This works best with old-style cathode ray tube (CRT) monitors that employed a raster scan of the image. Newer LCDs and other monitors may store the data, which arrives in serial form, and then illuminate the screen pixels all at once.

HAZARDS

There are no significant hazards with this demonstration other than inadvertently striking something with the stick or stumbling in the dark.

REFERENCES

1. H. J. Caulfield, Ed., *Handbook of Optical Holography*, Academic Press: New York (1979).

2. N. Abramson, *The Making and Evaluation of Holograms*, Academic Press: London (1981).

3. H. Spetzler, *Phys. Teach.* **24**, 80 (1986).

4. R. E. Latham, *Phys. Teach.* **24**, 395 (1986).

5. M. Linthwaite and C. Shimmens, *Phys. Teach.* **25**, 382 (1987).

6. F. Unterseher, J. Hansen, and B. Schlesinger, *The Holography Handbook*, Ross Books: Berkeley, CA (1987).

7. T. C. Altman, *Phys. Teach.* **26**, 223 (1988).

8. J. Iovine, *Homade Holograms: The Complete Guide to Inexpensive Do-It-Yourself Holography*, Tab Books: Blue Ridge Summit, PA (1990).

9. F. Wirth, *Phys. Teach.* **29**, 138 (1991).

10. D. W. Olson, *Phys. Teach.* **30**, 202 (1992).

11. H. Dittmann and W. B. Schneider, *Phys. Teach.* **30**, 244 (1992).

12. F. Defreitas, A. Rhody, and S. Michael, *Shoebox Holography*, Ross Books: Berkeley, CA (2000).

13. J. E. Kasper and S. A. Feller, *The Complete Hologram Book: How They Work and How to Make Them*, Dover: New York (2001).

6.12 Optical Illusions

Transparencies containing optical illusions projected on the wall or a screen illustrate the role of subjectivity in scientific experiments.

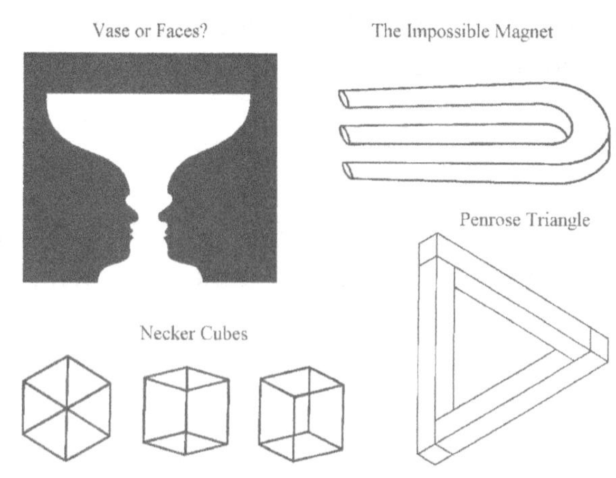

MATERIALS

- transparencies of optical illusions
- overhead projector or computer and LCD projector

PROCEDURE

Although optical illusions are often discussed in the context of psychology and art, you can also use them to illustrate principles of physics. Books and articles [1–15] provide numerous examples of optical illusions. You can photocopy many of these illusions onto transparencies and project them on a screen for viewing by a large audience or save them as computer graphics files and display them with an LCD projector. You can create impressive three-dimensional illusions, such as the Penrose (impossible) triangle shown in the figure, but you must generally view these images from a particular angle, and thus they are less suitable for large groups. To get a laugh, tell them "it is not an optical illusion; it just looks like one."

DISCUSSION

Frequently, detailed observation of a phenomenon reveals information not apparent and sometimes even contradictory to that gleaned from a casual observation. The result of a physical measurement often depends on the frame of reference of the observer as in the theory of relativity, and the act of making an observation can change the object observed as in the Heisenberg uncertainty principle. Everyone, but scientists in particular, must observe carefully and objectively and guard against the tendency to see in nature only what we expect to see.

HAZARDS

There are no significant hazards in the observation of these illusions.

REFERENCES

1. S. Luckiesh, *Visual Illusions: Their Causes, Characteristics and Applications*, Dover: New York (1965).
2. G. H. Fisher, *Perception and Psychophysics* **4**, 189 (1968).
3. F. Attneave, *Scientific American* **225**, 62 (Dec 1971).
4. L. Kettelkamp, *Tricks of Eye and Mind: The Story of Optical Illusion*, William Morrow and Company: New York (1974).
5. M. L. Teuber, *Scientific American* **231**, 90 (Jul 1974).
6. S. Simon, *The Optical Illusion Book*, Four Winds Press: New York (1976).
7. E. Lanners, Ed. (Translated and adapted by H. Norden), *Illusions*, Holt, Reinhart, and Winston: New York (1977).
8. M. Gardner, *Mathematical Magic Show*, Random House: New York (1977).
9. D. D. Hoffman, *Scientific American* **249**, 154 (Dec 1983).
10. J. Frederick, *Classical Illusions*, McPherson & Company: New Paltz, New York (1985).
11. S. Simon and C. Ftera, *Now You See It, Now You Don't: The Amazing World of Optical Illusions*, William Morrow & Company: New York (1998).
12. A. Seckel, *The Art of Optical Illusions*, Carlton Books: New York (2000).
13. A. Seckel, *The Great Book of Optical Illusions*, Firefly Books: Toronto (2002).
14. G. A. Sarcone and M. J. Waeber, *Dazzling Optical Illusions*, Sterling Publishing: New York (2002).
15. J. Slocum and J. Botermans, *Tricky Optical Illusion Puzzle*, Sterling Publishing: New York (2003).

6.13 Fractals

Transparencies or computer images containing fractals are projected on the wall or a screen.

MATERIALS

- string, piece of paper, and ball
- natural objects such as sticks, leaves, and rocks
- transparencies or computer images of fractals
- video camera and monitor (optional)

PROCEDURE

There are many ways to introduce fractals [1–7]. Point out that objects like a string are essentially one-dimensional, a piece of paper is two-dimensional, and a ball is three-dimensional. Then hold up a stick, and ask what its dimension is. Explain that when you view it from a large distance, it appears as a point (zero dimensions), but closer up, it looks one-dimensional, and then two-dimensional, and finally three-dimensional as you examine it more closely. Point out that objects in nature do not neatly fit into categories of integer dimension. After explaining that a sheet of paper is nearly two-dimensional, crumple it into a ball, and explain that it is now three-dimensional. Then slowly flatten it out, and ask at what point it becomes two-dimensional again. Explain that the dimension slowly changes from two to three as you again crumple it, and thus the concept of noninteger dimension is useful for describing such objects. However, the defining quality of a fractal is not its noninteger dimension, although most have that feature, but rather the fact that it has structure on all scales and contains infinitely many (perhaps approximate) copies of itself on ever smaller scales, called "self-similarity."

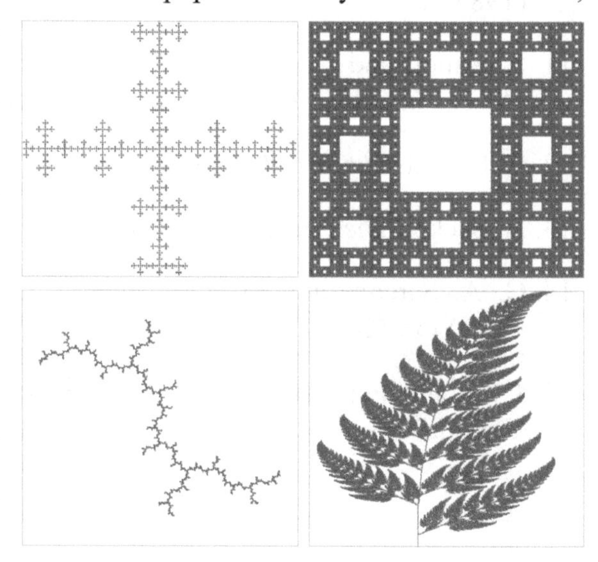

FRACTAL

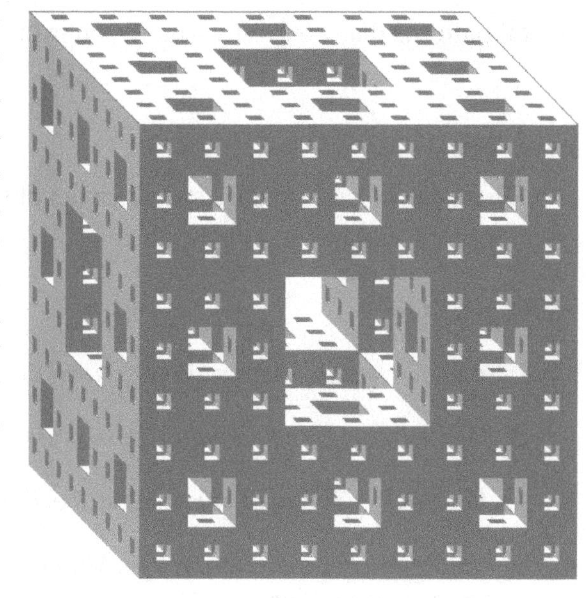

Other familiar examples of fractals are leaves (especially ferns), cracks in the sidewalk, coastlines and rivers, the pattern produced by a random walk (see section 1.21), and the electrical discharge from lightning or from a Tesla coil (see section 4.5). Fractals usually occur in the trajectory or orbit of a chaotic system, called a "strange attractor" [8]. You can produce fractals with simple computer programs or download them from many sites on the web and make an impressive slide show, bordering on artistic.

DISCUSSION

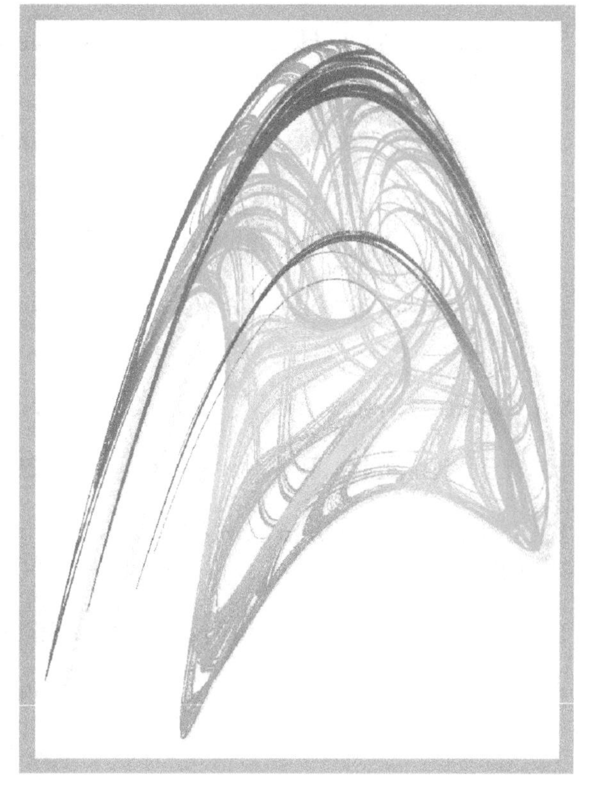

A fractal is a self-similar object with structure on all scales and usually a noninteger dimension. It may be exactly self-similar or only statistically self-similar. Natural objects are only approximately fractal since there is typically a limited range of scales over which they exhibit self-similarity, and the self-similarity is only approximate even within that range. Exactly self-similar objects are mathematical abstractions, never fully achieved in nature. Because of the fine-scale structure, they never become smooth, and hence the mathematics of calculus that requires smooth functions to define a derivative is of limited use.

You can understand the fractional dimension that characterizes most fractals in terms of their space-filling property [9]. The mass of a one-dimensional object like a string increases linearly with its size (length), the mass of a two-dimensional object like a sheet of paper increases with the square of its linear dimension, and the mass of a three-dimensional object like a ball increases with the cube of its linear size (diameter). The mass of square sheets of paper of various linear sizes similarly crumpled increases faster than the square of the diameter of the wad but slower than the cube of the diameter. A log-log plot of the mass versus the diameter would have a slope equal to the fractal dimension of the crumpled paper [10, 11]. In a similar way, you can determine the fractal dimension of a slice of bread [12].

A video camera aimed at a monitor to which it is connected can make interesting fractal patterns that evolve in time if you rotate the camera at an appropriate angle [13].

HAZARDS

There are no significant hazards in the observation of these fractals.

REFERENCES

1. J. Feder, *Fractals*, Plenum Press: New York (1982).
2. B. Mandelbrot, *The Fractal Geometry of Nature*, W. H. Freeman: New York (1983).
3. M. Barnsley, *Fractals Everywhere,* Academic Press: Boston (1988).
4. K. Falconer, *Fractal Geometry: Mathematical Foundations and Applications*, John Wiley & Sons: Chichester (1990).
5. M. O. Peitgen, H. Jurgens, and D. Saupe, *Chaos and Fractals: New Frontiers of Science*, Springer-Verlag: New York (1992).
6. V. Talanquer and G. Irazoque, *Phys. Teach.* **31**, 72 (1993).
7. M. Frame and B. Mandelbrot, *Fractals, Graphics & Mathematicsl Education*, The Mathematical Association of America: Washington (2002).
8. J. C. Sprott, *Strange Attractors, Creating Patterns in Chaos*, M&T Books: New York (1993).
9. K. Zembrowska and M. Kuzma, *Phys. Teach.* **40**, 470 (2002).
10. R. H. Ko and C. P. Bean, *Phys. Teach.* **29**, 78 (1991).
11. J. C. Sprott, *Chaos and Time-Series Analysis*, Oxford University Press: Oxford (2003).
12. D. H. Esbenshade, *Phys. Teach.* **29**, 236 (1991).
13. J. P. Crutchfield, *Physica D* **10**, 229 (1984).

Appendix A
Bibliography

Alley, M. *The Craft of Scientific Presentations: Critical Steps to Succeed and Critical Errors to Avoid*, Springer: New York (2003).

Amerongen, C. V. *The Way Things Work*, Simon and Schuster: New York (1967).

Arons, A. B. *A Guide to Introductory Physics Teaching*, John Wiley & Sons: New York (1990).

Berry, D. *A Potpourri of Physics Teaching Ideas*, American Association of Physics Teachers: College Park, MD (1987).

Blasi, R. C., Ed. *Physics Fun and Demonstrations with Professor Julius Sumner Miller*, Central Scientific Company: Franklin Park, IL (1974).

Bloomfield, L. A. *How Things Work: The Physics of Everyday Life* (2nd ed.), John Wiley & Sons: New York (2001).

Bohren, C. F. *Clouds in a Glass of Beer*, John Wiley & Sons: New York (1984).

Bolton, W. *Physics Experiments and Projects*, Pergamon Press: Oxford (1968).

Bord, D. J. and J. C. Sprott. *Great Ideas for Teaching Physics*, West Publishing Company: Saint Paul (1991).

Brown, R. J. *200 Illustrated Science Experiments for Children*, Tab Books: Blue Ridge Summit, PA (1987).

Brown, T. B., Ed. *Advanced Undergraduate Experiments in Physics*, Addison-Wesley Publishing Company: Reading, MA (1959).

Bulman, A. D. *Model-Making for Physicists*, Crowell Company: New York (1964).

Bulman, A. D. *Models for Experiments in Physics*, Crowell Company: New York (1966).

Carpenter, D. R. and R. B. Minnix. *The Dick and Rae Physics Demo Handbook*, Dick and Rae, Inc.: Lexington, VA (1993).

Chahrour, J. P. and A. H. Williams. *Flash! Bang! Pop! Fizz!: Exciting Science for Curious Minds*, Barrons Educational Books: Hauppauge, NY (2000).

Cherrier, F. *Fascinating Experiments in Physics*, Sterling Publishing Company: New York (1978).

Churchill, E. R., L. V. Loeschnig, and M. Mandell. *365 Super Science Experiments*, Sterling Publishing Company: New York (2001).

Conner, D. B., Ed. *A Potpourri of Physics Teaching Ideas*, American Association of Physics Teachers: College Park, MD (1987).

Cook, J. G. *The Thomas Edison Book of Easy and Incredible Experiments,* Dodd Mead: New York (1988).

Cunningham, J. and N. Herr. *Hands-On Physics Activities with Real-Life Applications*, The Center for Applied Educational Research in Education: West Nyack, NY (1994).

Doherty, P. and D. Rathjen, Eds. *The Exploratorium Science Snackbook*, Exploratorium: San Francisco (1991).

Downie, N. A. *Vacuum Bazookas, Electric Rainbow Jelly, and 27 Other Saturday Science Projects*, Princeton University Press: Princeton, NJ (2001).

Edge, R. D. *String and Sticky Tape Experiments*, American Association of Physics Teachers: College Park, MD (1987).

Ehrlich, R. *Turning the World Inside Out and 174 Other Simple Physics Demonstrations*, Princeton University Press: Princeton, NJ (1990).

Ehrlich, R. *Why Toast Lands Jelly-Side Down*, Princeton University Press: Princeton, NJ (1997).

Esler, W. K. *Modern Physics Experiments for the High School*, Parker Publishing Company: West Nyack, NY (1970).

Freier, G. D. and F. J. Anderson. *A Demonstration Handbook for Physics* (2nd ed.), American Association of Physics Teachers: College Park, MD (1981).

Gardner, M. *Entertaining Science Experiments with Everyday Objects,* Dover: New York (1981).

Gardner, R. *Science Experiments*, Franklin Watts: New York (1988).

Gardner, R. *More Ideas for Science Projects*, Franklin Watts: New York (1989).

Gardner, R. *Science Projects about Electricity and Magnetism*, Enslow Publishers, Inc.: Springfield, NJ (1994).

Gibbs, K. *The Resourceful Physics Teacher: 600 Ideas for Creative Teaching*, Institute of Physics Publishing: Bristol and Philadelphia (1999).

Goodman, D. S. *Optics Demonstrations with the Overhead Projector*, American Association of Physics Teachers: College Park, MD (2000).

Gottlieb, H. H. *Experiments for Physics Labs*, Analog Press: San Francisco, CA (1981).

Greenberg, L. H. *Discovery in Physics*, Saunders: Philadelphia (1968).

Handwerker, J. J. *Ready-to-Use Physical Science Activities*, Center for Applied Research in Education: West Nyack, NY (2000).

Herbert, D. *Mr. Wizard's Supermarket Science*, Random House: New York (1980).

Herbert, D. and H. Ruchlis. *Mr. Wizard's 400 Experiments in Science*, Revised Edition, Booklab: North Bergen, NJ (1983).

Hewitt, P. *Physics for Phun*, American Association of Physics Teachers: College Park, MD (1990).

Hilton, W. A. Physics *Demonstration Experiments at William Jewell College*, William Jewell College: Liberty, MO (1971).

Ingersoll, L. R., M. J. Martin, and T. A. Rouse. *Experiments in Physics*, McGraw-Hill: New York (1966).

Jewett, J. W. *Physics Begins with an M: Mysteries, Magic, and Myth*, Allyn & Bacon: Des Moines, IA (1996).

Jones, E. G. *Physics Demonstrations and Experiments for High School*, Physics Department, Mississippi State University: Mississippi State, MS (1984).

Joseph, A., P. F. Brandwein, E. Morholt, H. Pollack, and J. F. Castka. *A Sourcebook for the Physical Sciences*, Harcourt, Brace and World: New York (1961).

Kardos, T. *75 Easy Physics Demonstrations*, J. Weston Walch: Portland, ME (1996).

Kardos, T. *Easy Science Demos & Labs*, J. Weston Walch: Portland, ME (2003).

Kenda, M. and P. Williams. *Science Wizardry for Kids*, Barrons: Hong Kong (1992).

Kinsman, E. M. and C. Waters. *Unique Science: Demonstrations & Laboratories for the Physics Instructor*, Kinsman Physics: Syracuse, NY (1991).

Liem, T. K. *Invitations to Science Inquiry*, Ginn Press: Lexington MA (1981).

Lunetta, V. N. and S. Novick. *Inquiring and Problem-Solving in the Physical Sciences: A Sourcebook*, Kendall/Hunt: Dubuque, IA (1982).

Lynde, C. *Science Experiences with Inexpensive Equipment*, International Textbook: Scranton, PA (1939).

Macaulay, D. *The New Way Things Work*, Houghton Mifflin: Boston (1998).

Mamola, K. C. *Apparatus for Teaching Physics*, American Association of Physics Teachers: College Park, MD (1998).

Mark, H. M. and N. T. Olson. *Experiments in Modern Physics*, McGraw-Hill: New York (1936).

McComb, G. *Gadgeteer's Goldmine! 55 Space-Age Projects*. TAB Books, McGraw-Hill: Blue Ridge Summit, PA (1990).

McCullough, J. and R. McCullough. *The Role of Toys in Teaching Physics*, American Association of Physics Teachers: College Park, MD (1995).

Meiners, H. F., Ed. *Physics Demonstration Experiments*, Vols. 1 and 2, The Ronald Press Company: New York (1970).

Meiners, H. F., W. Eppenstein, R. A. Oliva, and T. Shannon. *Laboratory Physics*, John Wiley & Sons: New York (1987).

Melissinos, A. C. *Experiments in Modern Physics*, Academic Press: New York (1966).

Morse, R. *Teaching About Electrostatics*, American Association of Physics Teachers: College Park, MD (1992).

Nokes, M. C. *Demonstrations in Modern Physics*, William Heinemann Ltd.: Melbourne (1952).

Noll, C. K. *The Ben Franklin Book of Easy and Incredible Experiments: A Franklin Institute Science Museum Book,* John Wiley & Sons: New York (1995).

Nye, B. *Bill Nye The Science Guy,* Hyperion Press: Westport, CT (2000).

Pepper, J. H. *The Boy's Playbook of Science*, Routledge, Warne, and Routledge: London (1860).

Pizzo, J., Ed. *Interactive Physics Demonstrations*, American Association of Physics Teachers: College Park, MD (2001).

Potter, J. *Science in Seconds for Kids: Over 100 Experiments You Can Do in Ten Minutes or Less,* John Wiley & Sons: New York (1995).

Provenzo, E. F. *47 Easy-to-Do Classic Science Experiments,* Dover: New York (1989).

Rathjen, D. and P. Doherty. *Square Wheels and Other Easy-to-build Hands-on Science Activities*, Exploratorium Teacher Institute: San Francisco (2002).

Richards, K. *Science Magic with Physics*, Arco Publishing Company: New York (1975).

Robinson, H. A., Ed. *Lecture Demonstrations in Physics*, American Institute of Physics: New York (1963).

Rodecker, S. and M. Quon-Warner. *Laboratory Experiments and Activities in Physical Science*, Spectrum Publications: New York (1996).

Shakhashiri, B. Z. *Chemical Demonstrations*, The University of Wisconsin Press: Madison, WI (1983).

Shmaefsky, B. R., Ed. *Favorite Demonstrations for College Science*, National Science Teachers Association Press: Washington (2004).

Slater, T. F. and M. Zeilik, Ed. *Insights Into the Universe: Effective Ways to Teach Astronomy*, American Association of Physics Teachers: College Park, MD (2003).

Sutton, R. M., Ed. *Demonstration Experiments in Physics*, McGraw-Hill: New York (1938).

Sweezy, K. *After-Dinner Science*, McGraw-Hill: New York (1948).

Sweezy, K. *Science Magic*, McGraw-Hill: New York (1952).

Taylor, C. *The Art and Science of Lecture Demonstration*, American Institute of Physics: New York (1988).

Taylor, B., J. Poth, and D. Portman. *Teaching Physics with Toys*, Terrific Science Press: Middletown, OH (1995).

Van Cleave, J. P. *Teaching the Fun of Physics*, Prentice-Hall: New York (1985).

Walker, J., B. Carroll, J. A. Davis, and R. Berg. *The Video Encyclopedia of Physics Demonstrations*, Frey Scientific: Mansfield, OH (2002).

Walker, J. W. *The Flying Circus of Physics—With Answers*, Interscience Publishers, John Wiley & Sons: New York (1977).

Walker, J. W. *The Physics of Everyday Phenomena*, W. H. Freeman: San Francisco (1979).

Wendt, G. *700 Science Experiments for Everyone*, Doubleday: Garden City, NY (1958).

Wood, R. W. *Physics for Kids*, McGraw-Hill: Blue Ridge Summit, PA (1989).

Wood, R. W. *The McGraw-Hill Big Book of Science Experiments,* McGraw-Hill: Blue Ridge Summit, PA (1999).

Appendix B
Selected Vendors of Scientific Demonstration Equipment

American 3B Scientific
2189 Flintstone Drive, Suite O
Tucker, GA 30084
770-492-9111
http://www.a3bs.com/

American Science and Surplus
P.O. Box 1030
3605 Howard Street
Skokie, IL 60076
888-724-7587
http://www.sciplus.com/

American Scientific
6450 Fiesta Drive
Columbus, OH 43235
888-490-9002
http://www.american-scientific.com/

Arbor Scientific
P. O. Box 2750
Ann Arbor, MI 48106-2750
734-477-9370
http://www.arborsci.com/

Astronomical Society of the Pacific
390 Ashton Avenue
San Francisco, CA 94112-1722
415-337-1100
http://www.astrosociety.org/

Carolina Biological Supply Company
P. O. Box 6010
2700 York Road
Burlington, NC 27216-6010
336-584-0381
http://www.carolina.com/

Cole-Parmer
625 East Bunker Court
Vernon Hills, IL 60061
800-323-4340
http://www.coleparmer.com/

CoolStuffCheap
195 Libbey Industrial Parkway #1
Weymouth, MA 02189
561-223-2458
http://www.coolstuffcheap.com/

CPO Science
26 Howley Street, 3rd Floor
Peabody, MA 01960
800-932-5227
http://cposcience.com/

Edmund Optics
101 East Gloucester Pike
Barrington, NJ 08007-1380
800-363-1992
http://www.edmundoptics.com/

Educational Innovations
5 Francis J. Clarke Circle
Bethel, CT 06801
203-748-3224
http://www.teachersource.com/

Estes-Cox Corporation
P. O. Box 227
1295 H Street
Penrose, CO 81240-0227
719-372-6565
http://www.estesrockets.com/

Fisher Science Education
4500 Turnberry Drive
Hanover Park, IL 60133
800-766-7000
http://www.fishersci.com/

Forcefield
2606 West Vine Drive
Fort Collins, CO 80521
970-484-7257
http://www.wondermagnet.com/

Frey Scientific
P. O. Box 3000
Nashua, NH 03061-3000
800-225-3739
http://www.freyscientific.com/

GDJ Incorporated
7585 Tyler Boulevard
Mentor, OH 44060
440-975-0258
http://www.gdjinc.com/

H&R Company
353 Crider Avenue
Moorestown, NJ 08057
856-802-0422
http://www.herbach.com/

Hampden Engineering Corporation
P. O. Box 563
99 Shaker Road
East Longmeadow, MA 01028
413-525-3981
http://www.hampden.com/

HEMCO Corporation
711 S. Powell Road
Independence, MO 64056
816-796-2900
http://hemcocorp.com/

Industrial Fiber Optics
1725 West 1st Street
Tempe, AZ 85281-7622
480-804-1227
http://i-fiberoptics.com/

Klinger Educational Products Corporation
112-19 14th Road
College Point, NY 11356
718-461-1822
http://www.klingereducational.com/

Labconco Corporation
8811 Prospect Avenue
Kansas City, MO 64132-2696
816-333-8811
http://www.labconco.com/

Lafayette Instrument Company
P. O. Box 5729
3700 Sagamore Parkway North
Lafayette, IN 47903
765-423-1505
http://www.lafayetteinstrument.com/

Lenox Laser
12530 Manor Road
Glen Arm, MD 21057
410-592-3106
http://www.lenoxlaser.com/

Merlan Scientific
234 Matheson Boulevard E
Mississauga, ON Canada L4Z 1X1
800-387-2474
http://www.merlan.ca/

Nasco
P. O. Box 901
901 Janesville Avenue
Fort Atkinson, WI 53538-0901
800-372-1236
http://www.enasco.com/

PASCO Scientific
10101 Foothills Boulevard
Roseville, CA 95747-7100
916-786-3800
http://www.pasco.com/

Sargent-Welch
P. O. Box 92912
Rochester, NY 14692-9012
800-727-4368
https://www.sargentwelch.com/

Schoolmasters Science
745 State Circle
Ann Arbor, MI 48108
800-654-4321
http://schoolmasters.com/

Science First
95 Botsford Place
Buffalo, NY 14216
904-225-5558
http://www.sciencefirst.com/

Scientifics
532 Main Street
Tonawanda, NY 14150
800-818-4955
http://www.scientificsonline.com/

Spectra-Physics
3635 Peterson Way
Santa Clara, CA 95054
408-980-4300
http://www.spectra-physics.com/

Spectrum Techniques
106 Union Valley Road
Oak Ridge, TN 37830
865-482-9937
http://www.spectrumtechniques.com/

Starlab
86475 Gene Lasserre Blvd.
Yulee, FL 32097
904-225-5558
http://starlab.com/

Supersaturated Environments
P. O. Box 55252
Madison, WI 53705
608-238-5068
http://www.cloudchambers.com/

Surplus Shed
1050 Maidencreek Road
Fleetwood, PA 19522
610-926-9226
http://surplusshed.com/

Teachspin, Incorporated
Tri-Main Center, Suite 409
2495 Main Street
Buffalo, NY 14214
716-885-4701
http://www.teachspin.com/

Team Labs Corporation
6859 North Foothills Highway, Building D200
Boulder, CO 80302
303-541-9001
http://www.teamlabs.com/

TEL-Atomic, Incorporated
P. O. Box 924
Jackson, MI 49204-0924
517-783-3039
http://www.telatomic.com/

The Science Source
86475 Gene Lasserre Blvd.
Yulee, FL 32097
904-225-5558
http://www.thesciencesource.com/

True Mirror Company, Incorporated
43 East First Street
New York, NY 10003
212-614-6636
http://www.truemirror.com/

Vernier Software & Technology
13979 S.W. Millikan Way
Beaverton, OR 97005-2886
503-277-2299
http://www.vernier.com/

Ward's Science
P. O. Box 92912
5100 West Henrietta Road
Rochester, NY 14692-9102
800-962-2660
https://www.wardsci.com/

Appendix C
Selected Vendors of Audiovisual and Computer Materials

Agency for Instructional Technology
8111 N. Lee Paul Road
Bloomington, IN 47404-7916
812-339-2203
http://www.ait.net/

American Association of Physics Teachers
One Physics Ellipse
College Park, MD 20740-3845
301-209-3311
http://www.aapt.org/

BearEdu Technologies
2301 Live Oak Circle
Round Rock, TX 78681
512-914-0603
http://bearedu.com/

Design Simulation Technologies, Incorporation
43311 Joy Road, #237
Canton, MI 48187
415-259-4207
http://workingmodel.design-simulation.com/

Destination Education
4910 S. 75th Street
Lincoln, NE 68516
402-435-0110
http://shopgpn.com/

Discovery Store Education
700 Indian Springs Drive
Lancaster, PA 17601
888-892-3484
http://store.discoveryeducation.com/

EME Corporation
P. O. Box 1949
581 Central Parkway
Stuart, FL 34995
772-219-2206
http://www.emescience.com/

Engineering Software
P. O. Box 1180
Germantown, MD 20875
301-540-3605
http://www.engineering-4e.com/

Fable Multimedia Ltd.
109-113 Royal Avenue
Belfast BT1 1FF
Northern Ireland, UK
+44(0)-2890-320-736
http://www.fable.co.uk/

Films Media Group
132 West 31st Street, 17th Floor
New York, NY 10001
800-322-8755
http://www.films.com/

Hawkhill Associates, Incorporated
125 East Gilman Street
Madison, WI 53703
608-467-7003
http://www.hawkhill.com/

Insight Media, Incorporated
350 7th Avenue, Suite 1100
New York, NY 10001
212-721-6316
http://www.insight-media.com/

Jones & Bartlett Learning
5 Wall Street
Burlington, MA 01803
978-443-5000
http://www.jblearning.com/

MMI Corporation
P. O. Box 19907
2950 Wyman Parkway
Baltimore, MD 21211
410-366-1222
http://www.mmicorporation.com/

Perfection Learning
P. O. Box 500
1000 North Second Avenue
Logan, IA 51546
712-644-2831
http://www.perfectionlearning.com/

Physics Curriculum and Instruction
22585 Woodhill Drive
Lakeville, MN 55044
952-461-3470
http://www.physicscurriculum.com/

Queue, Incorporated
80 Hathaway Drive
Stratford, CT 06615
800-2322224
http://qworkbooks.com/

Teacher's Media Company
5815 Live Oak Parkway, Suite 2-B
Norcross, GA 30093-1700
800-451-5226
http://www.schoolmediaassociates.com/

U V Process Supply, Incorporated
1229 West Cortland Street
Chicago, IL 60614-4805
773-248-0099
http://www.uvprocess.com/

Variable Symbols, Incorporated
356 Bush Street
Mountain View, CA 94041
650-966-8999
http://www.variablesymbols.com/

WebAssign
1791 Varsity Drive, Suite 200
Raleigh, NC 27606
919-829-8181
https://webassign.com/

Ztek Company
P. O. Box 967
Lexington, KY 40588-0967
859-281-1611
http://www.ztek.com/

Appendix D
Other Related Materials

Videos

Video DVDs of "The Wonders of Physics" containing most of the demonstrations described in this book performed before a live audience of the general public are available for $25 each postpaid (in the United States) from:

The Wonders of Physics
University of Wisconsin
1150 University Avenue
Madison, WI 53706
608-262-2927
http://sprott.physics.wisc.edu/wop.htm

The Wonders of Physics Lecture Kit

This kit of materials describes "The Wonders of Physics" program in detail. It includes a recent videotape of "The Wonders of Physics," the *Physics Demonstrations* software (see below), a book explaining the motivational and logistical considerations, and samples of handouts and publicity materials. It is an aid for physics teachers and physicists who want to start similar programs or enhance already existing ones. It is available for $90 postpaid from:

The Wonders of Physics
University of Wisconsin
1150 University Avenue
Madison, WI 53706
608-262-2927
http://sprott.physics.wisc.edu/wop.htm

Computer Animations

Computer simulations of many of the demonstrations as listed below are available for electronic delivery. A second, more advanced package of programs, *Chaos Demonstrations*, is also available. The programs are suitable for use as part of a lecture/demonstration or as museum displays. Additional information about *Physics Demonstrations* and *Chaos Demonstrations* is available from the author.

Physics Demonstrations
Computer Software Version 1.22

Motion:
- Reaction Time (§1.2)
- Ballistics Car (§1.3)
- The Monkey and the Coconut (§1.4)
- Bowling Ball Pendulum (§1.10)
- Wilberforce Pendulum (§1.19)

Sound:
- Wave Speed on a Rope (§3.1)
- Flame Pipe (§3.6)
- Oscilloscope Waveforms (§3.7)
- Breaking a Beaker with Sound (§3.9)
- Doppler Effect (§3.5)

Chaos Demonstrations
Computer Software Version 3.2

- Driven Pendulum
- Nonlinear Oscillator
- Duffing Oscillator
- Van der Pol Equation
- Three-Body Problem
- Magnetic Quadrupole
- Lorenz Attractor
- Rössler Attractor
- One-Dimensional Maps
- Predator-Prey
- Chirikov Map
- Hénon Map
- Strange Attractors
- Mandelbrot Set
- Julia Sets
- Diffusion
- Noise
- Deterministic Fractals
- Random Fractals
- Iterated Function Systems
- Coupled-Map Lattice
- Mixing

- Percolation
- Cellular Automata
- Game of Life
- Anaglyph

Index